M. Conn   C. Kordon
Y. Christen (Eds.)

Insights into Receptor Function and New Drug Development Targets

With 58 Figures and 14 Tables

 Springer

Conn, Michael, Ph.D.
Divisions of Neuroscience and Reproductive Biology
Departments of Physiology and Pharmacology
and Cell and Developmental Biology
505 NW 185th Avenue
Oregon National Primate Research Center
Oregon Health and Science University
Beaverton, Oregon 97006
USA
email: connm@ohsu.edu

Kordon, Claude, Ph.D.
Institut Necker
156, rue de Vaugirard
75015 Paris
France
e-mail: kordon@necker.fr

Christen, Yves, Ph.D.
Fondation IPSEN
Pour la Recherche Thérapeutique
24, rue Erlanger
75781 Paris Cedex 16
France
e-mail: yves.christen@ipsen.com

ISBN-10 3-540-34446-2 Springer Berlin Heidelberg New York
ISBN-13 978-3-540-34446-9 Springer Berlin Heidelberg New York

Cataloging-in-Publication Data applied for Bibliographic information published by Die Deutsche Bibliothek
Die Deutsche Bibliothek lists this publication in the Deutsche Nationalbibliografie; detailed bibliographic data is available in the Internet at <http://dnb.ddb.de>.

Springer is a part of Springer Science+Business Media
springer.com

© Springer-Verlag Berlin Heidelberg 2006
Printed in Germany

The use of general descriptive names, registered names, trademarks, etc. in this publication does not imply, even in the absence of a specific statement, that such names are exempt from the relevant protective laws and regulations and therefore free for general use.

Product liability: The publishers cannot guarantee the accuracy of any information about dosage and application contained in this book. In every individual case the user must check such information by consulting the relevant literature.

Cover design: *design & production*, Heidelberg, Germany
Typesetting and production: LE-TEX Jelonek, Schmidt & Vöckler GbR, Leipzig, Germany
Printed on acid-free paper   27/3100/YL   5 4 3 2 1 0
SPIN 11757955

Introduction

> *"You cannot suppose that atoms of the same shape are entering our nostrils when stinking corpses are roasting as when the stage is freshly sprinkled with saffron of Cilicia and a near-by alter exhales the perfumes of the Orient. You cannot attribute the same composition to sights that feast the eye with colour and to those that make it smart and weep or that appear loathsome and repulsive through sheer ugliness. Nothing that gratifies the sense is ever without a certain smoothness of the constituent atoms. Whatever, on the other hand, is painful and harsh is characterized by a certain roughness of matter"*
>
> (Lucretius, *Nature of the Universe, Book II*).

The origin of the concept of the ligand-receptor interaction clearly stretches back to the times of our philosophical forbearers, but it is only in the recent part of the current century that we have aggressively harnessed the power of our understanding of structure-activity relationships in order to guide the process of drug design. Still more recently came the approach of chemical alteration of drug chemistry in order to modify their pharmacological properties.

We are now on the verge of viewing effector molecules and other regulatory sites as therapeutic targets for the amelioration of human and animal disease. The recognition, for example, that mutant proteins are frequently misrouted molecules, rather than functionally defective ones, changes our approach to "inborn errors of metabolism" and offers new approaches for pharmacological discovery, based on rescue of receptors, ion channels and enzymes with pharmacoperones. Ion channels, regulators of G-protein signaling and enzymes engaged in regulation, now present opportunities for drug development.

The state of our art also benefits by the availability of superior tools that allow measurement of interactions and afford unprecedented insight into the biomolecular interactions that present novel approaches to drug design.

It was in this spirit of excitement that this colloquium was organized during early 2004, at discussions held in San Diego and Lisbon. The result of these plans brought together leaders, from many countries and academic disciplines, to address novel sites of pharmacologic intervention with a common view of improving drug design by diversifying the approach, during the Colloque Médecine et Recherche held in Paris on December 5, 2005.

The organizers of this meeting would like to thank the presenters and authors of the articles in this volume, along with the expert staff at Fondation IPSEN, for their timely submission and the processing of articles respectively, which enabled production of a current volume. We hope that readers who unable to attend the meeting will take some of the excitement of the scientific interaction from this volume.

Paris, December 2005

Claude Kordon
Yves Christen
Michael Conn

Table of Contents

List of Contributors

Abramowitz, Joel
Laboratory of Signal Transduction, Division of Intramural Research,
National Institute of Environmental Health Sciences, National Institutes of Health,
Department of Health and Human Services, 111 T.W. Alexander Drive,
Research Triangle Park, North Carolina 27709, USA

Beaudet, Alain
Montréal Neurological Institute, Fonds de la Recherche en Santé Québec,
500, Rue Sherboocke Ouest, Bureau 800, Université McGill, Montréal, Quebec H3A 3C6,
Canada, abeaudet@frsq.gouv.qc.ca

Berlan, Michel
Unité de Recherches sur les Obésités, Université Paul Sabatier,
Institut Louis Bugnard IFR31, 31432 Toulouse, France

Blahos, Jaroslav
Dept. of Molecular Pharmacology, Institute of Experimental Medicine,
Academy of Science of Czech Rebublic, Videnska 1083, 14220 Prague 4, Czech Republic

Binet, Virginie
Institute of Functional Genomic, Department of Molecular Pharmacology,
CNRS UMR5203, INSERM U661, Université de Montpellier 1,
Université de Montpellier 2, Montpellier, France

Birnbaumer, Lutz
Laboratory of Signal Transduction, Division of Intramural Research,
National Institute of Environmental Health Sciences, National Institutes of Health,
Department of Health and Human Services, 111 T.W. Alexander Drive,
Research Triangle Park, North Carolina 27709, USA, birnbau1@niehs.nih.gov

Blahos, Jaroslav
Dept. of Molecular Pharmacology, Institute of Experimental Medicine,
Academy of Science of Czech Rebublic, Videnska 1083, 14220 Prague 4, Czech Republic

Bodineau, Laurence
Inserm U691, Collège de France, 11, Place Marcelin Berthelot, 75005 Paris, France

Brock, Carsten
Institute of Functional Genomic, Department of Molecular Pharmacology,
CNRS UMR5203, INSERM U661, Université de Montpellier 1,
Université de Montpellier 2, Montpellier, France

Brothers, Shaun P.
Division of Neuroscience, Oregon National Primate Research Center,
Department of Physiology and Pharmacology, Oregon Health and Science University,
505 NW 185th Avenue, Beaverton, Oregon 97006, USA

Cescato, Renzo
Division of Cell Biology and Experimental Cancer Research, Institute of Pathology,
University of Berne, Murtenstrasse 31 P.O. Box 62, 3010 Berne, Switzerland

Clement, Karine
Inserm "Avenir", Université Pierre et Marie Curie – Paris 6,
IFR58, CHRU Pitié Salpétrière –, Service de Nutrition-Hôtel-Dieu,
Place du Parvis Notre-Dame, 75004 Paris, France, karine.clement@htd.ap-hop-paris.fr

Chartrel, Nicolas
Laboratoire d'Endocrinologie Moléculaire, Inserm U413, Université de Rouen,
Place Émile Blondel, 76821 Mont Saint Aignan Cedex, France

Conn, Michael
Divisions of Neuroscience and Reproductive Biology,
Departments of Physiology and Pharmacology and Cell and Developmental, Biology,
505 NW 185th Avenue, Oregon National Primate Research Center,
Oregon Health and Science University, Beaverton, Oregon 97006, USA,
connm@ohsu.edu

Costagliola, Sabine
IRIBHM, Université Libre de Bruxelles, Campus Erasme, 808, Route de Lennik,
1070 Bruxelles, Belgique

Crampes, François
Unité de Recherches sur les Obésités, Université Paul Sabatier,
Institut Louis Bugnard IFR31, 31432 Toulouse, France

De Mota, Nadia
Inserm U691, Collège de France, 11, Place Marcelin Berthelot, 75005 Paris, France

DiGruccio, Michael
The Clayton Foundation Laboratories for Peptide Biology,
The Salk Institute for Biological Studies, 10010 N. Torrey Pines Road, La Jolla 92037,
USA

Dubern, Béatrice
Inserm "Avenir", Université Pierre et Marie Curie – Paris 6, IFR58,
CHRU Pitié Salpétrière –, Service de Nutrition-Hôtel-Dieu,
Place du Parvis Notre-Dame, 75004 Paris, France

El Messari, Said
Inserm U691, Collège de France, 11, Place Marcelin Berthelot, 75005 Paris, France,
said.el-messari@college-de-france.fr

Erchegyi, Judit
The Clayton Foundation Laboratories for Peptide Biology,
The Salk Institute for Biological Studies, 10010 N. Torrey Pines Road, La Jolla 92037,
USA

Eltschinger, Véronique
Division of Cell Biology and Experimental Cancer Research, Institute of Pathology,
University of Berne, Murtenstrasse 31 P.O. Box 62, 3010 Berne, Switzerland

Fischer, Oliver
Max-Planck-Institute of Biochemistry, Department of Molecular Biology,
Am Klopferspitz 18, 82152 Martinsried, Germany

Frugière, Alain
Inserm U691, Collège de France, 11, Place Marcelin Berthelot, 75005 Paris, France

Galitzky, Jean
Unité de Recherches sur les Obésités, Université Paul Sabatier,
Institut Louis Bugnard IFR31, 31432 Toulouse, France

Goudet, Cyril
Institute of Functional Genomic, Department of Molecular Pharmacology,
CNRS UMR5203, INSERM U661, Université de Montpellier 1,
Université de Montpellier 2, Montpellier, France

Grace, Christy R.
Structural Biology Laboratory, The Salk Institute, 10010 N. Torrey Pines Road,
La Jolla 92037, USA

Gschwind, Andreas
Max-Planck-Institute of Biochemistry, Department of Molecular Biology,
Am Klopferspitz 18, 82152 Martinsried, Germany

Gulyas, Jozsef
The Clayton Foundation Laboratories for Peptide Biology,
The Salk Institute for Biological Studies, 10010 N. Torrey Pines Road,
La Jolla 92037, USA

Hlavackova, Veronika
Dept. of Molecular Pharmacology, Institute of Experimental Medicine,
Academy of Science of Czech Rebublic,
Videnska 1083, 14220 Prague 4, Czech Republic

Hus-Citharel, Annette
Inserm U691, Collège de France, 11, Place Marcelin Berthelot, 75005 Paris, France,
annette.hus-citharel@college-de-france.fr

Iturrioz, Xavier
Inserm U691, Collège de France, 11, Place Marcelin Berthelot, 75005 Paris, France,
xavier.iturrioz@college-de-france.fr

Izzo, Angelo
Department of Experimental Pharmacology, University of Naples Federico II,
via D Montesano 49, 80131 Naples, Italy, aaizzo@unina.it

Kniazeff, Julie
Institute of Functional Genomic, Department of Molecular Pharmacology,
CNRS UMR5203, INSERM U661, Université de Montpellier 1,
Université de Montpellier 2, Montpellier, France

Koerber, Steven C.
The Clayton Foundation Laboratories for Peptide Biology,
The Salk Institute for Biological Studies,
10010 N. Torrey Pines Road, La Jolla 92037, USA

Kordon, Claude
Institut Necker, 156, Rue de Vaugirard, 75015 Paris, France, kordon@necker.fr

Lafontan, Max
IFR31 – Institut Louis Bugnard, Obesity Research Unit – Unité Inserm UPS – U586,
Université Paul Sabatier – Institut Louis Bugnard IFR31, BP84226, 31432 Toulouse,
France, max.lafontan@toulouse.inserm.fr

Liao, Yanhong
Laboratory of Signal Transduction, Division of Intramural Research,
National Institute of Environmental Health Sciences, National Institutes of Health,
Department of Health and Human Services, 111 T.W. Alexander Drive,
Research Triangle Park, North Carolina 27709, USA

Llorens-Cortes, Catherine
Inserm U691 – Collège de France, 11, Place Marcelin Berthelot, 75231 Paris Cedex 05,
France, c.llorens-cortes@college-de-france.fr

Lubrano, Cécile
Inserm "Avenir", Université Pierre et Marie Curie – Paris 6, IFR58,
CHRU Pitié Salpétrière –, Service de Nutrition-Hôtel-Dieu,
Place du Parvis Notre-Dame, 75004 Paris, France

Maurel, Damien
Institute of Functional Genomic, Department of Molecular Pharmacology,
CNRS UMR5203, INSERM U661, Université de Montpellier 1,
Université de Montpellier 2, Montpellier, France

Melmed, Shlomo
Cedars-Sinai Medical Center, Division of Endocrinology and Metabolism,
8700 Beverly Blvd., Room 2015, Los Angeles, CA 90048, USA, melmeds@cshs.org

Moos, Françoise
Institut F. Magendie Bx-2, Neurobiologie Intégrative, FRE 2723 CNRS, UMR 1244 INRA,
Université de Bordeaux 2, 33076 Bordeaux Cedex, France, fmoos@bordeaux.inra.fr

Moro, Cédric
Unité de Recherches sur les Obésités, Université Paul Sabatier,
Institut Louis Bugnard IFR31, 31432 Toulouse, France

Neubig, Richard R.
University of Michigan, Department of Pharmacology, 1301 MSRB III,
Ann Arbor, MI 48109-0632, USA, RNeubig@umich.edu

Pardo, Leonardo
Laboratori de Medicina Computacional, Unitat de Bioestadística, Facultat de Medicina,
Universitat Autònoma de Barcelona, 08193 Bellaterra, Spain

Perrin, Marilyn
The Clayton Foundation Laboratories for Peptide Biology,
The Salk Institute for Biological Studies, 10010 N. Torrey Pines Road, La Jolla 92037,
USA, perrin@salk.edu

Pin, Jean-Philippe
Institut de Génomique Fonctionelle –, Département de Pharmacologie Moléculaire,
UMR 5203 CNRS – U661 Inserm, Université de Montpellier I & II, 141,
Rue de la Cardonille, 34094 Montpellier Cedex 05, France, jean-philippe.pin@igf.cnrs.fr

Prezeau, Laurent
Institute of Functional Genomic, Department of Molecular Pharmacology,
CNRS UMR5203, INSERM U661, Université de Montpellier 1,
Université de Montpellier 2, Montpellier, France

Reaux-Le Goazigo, Annabelle
Inserm U691, Collège de France, 11, Place Marcelin Berthelot, 75005 Paris, France,
areaux@infobiogen.fr

Reubi, Jean-Claude
Division of Cell Biology and Experimental Cancer Research,
Institute of Pathology, University of Berne, Murtenstrasse 31 P.O. Box 62, 3010 Berne,
Switzerland., reubi@patho.unibe.ch

Riek, Roland
Structural Biology Laboratory, The Salk Institute, 10010 N. Torrey Pines Road,
La Jolla 92037, USA

Rondard, Philippe
Institute of Functional Genomic, Department of Molecular Pharmacology,
CNRS UMR5203, INSERM U661, Université de Montpellier 1,
Université de Montpellier 2, Montpellier, France

Rivier, Jean
The Clayton Foundation Laboratories for Peptide Biology,
The Salk Institute for Biological Studies, 10010 N. Torrey Pines Road, La Jolla 92037,
USA, jrivier@salk.edu

Rivier, Catherine
The Clayton Foundation Laboratories for Peptide Biology,
The Salk Institute for Biological Studies, 10010 N. Torrey Pines Road,
La Jolla 92037, USA

Sengenes, Coralie
Unité de Recherches sur les Obésités, Université Paul Sabatier,
Institut Louis Bugnard IFR31, 31432 Toulouse, France

Spiegel, Allen M.
National Institute of Diabetes, Digestive and Kidney Diseases, Bldg. 31, Rm. 9A/52,
National Institute of Health, Bethesda, MD 20892, USA, spiegela@extra.niddk.nih.gov

Ullrich, Axel
Department of Molecular Biology, Max Planck Institute of Biochemistry,
Am Klopferspitz 18a,, 82152 Martinsried, Munich, Germany, ullrich@biochem.mpg.de

Vale, Wylie
The Clayton Foundation Laboratories for Peptide Biology,
The Salk Institute for Biological Studies, 10010 N. Torrey Pines Road, La Jolla 92037,
USA, vale@salk.edu

Vassart, Gilbert
IRIBHM, Service de Génétique Médicale, Université Libre de Bruxelles,
Campus Erasme, 808 route de Lennik, 1070 Bruxelles, Belgique, gvassart@ulb.ac.be

Vaudry, Hubert
Laboratoire d'Endocrinologie Moléculaire, Inserm U413, Université de Rouen,
Place Émile Blondel, 76821 Mont Saint Aignan Cedex, France,
hubert.vaudry@univ-rouen.fr

Waser, Beatrice
Division of Cell Biology and Experimental Cancer Research, Institute of Pathology,
University of Berne, Murtenstrasse 31 P.O. Box 62, 3010 Berne, Switzerland

Yildirim, Eda
Laboratory of Signal Transduction, Division of Intramural Research,
National Institute of Environmental Health Sciences, National Institutes of Health,
Department of Health and Human Services, 111 T.W. Alexander Drive,
Research Triangle Park, North Carolina 27709, USA

Molecular and functional diversity of the TRPC family of ion channels.
TRPC channels and their role in ROCE/SOCE

Lutz Birnbaumer[1], *Eda Yildirim*[1], *Yanhong Liao*[1], and *Joel Abramowitz*[1]

TRP channels – An unexpectedly large family of largely non-selective cation channels with an unexpectedly wide spectrum of physiological roles

The Drosophila *trp* mutation is responsible for the phenotype called "transient receptor potential", an alteration of the fly's electrorentinogram in which its sustained phase is missing (Pak et al. 1970; Hotta and Benzer 1970). The responsible gene was cloned in 1989 (Montell and Rubin 1989). Its amino acid sequence predicted a protein with eight hydrophobic segments that could potentially form transmembrane segments. Purification and cloning of a calmodulin-binding protein from Drosophila heads showed it to be a homologue of *trp*. It received the name *trp-like* or *trpl* (Phillips et al. 1992). Its discoverers highlighted the existence of limited sequence similarities between *trp/trpl* and voltage-sensitive Na^+ and Ca^{2+} channels. Expression of *trpl* in silkworm cells of *Spodoptera frugiperda* (Sf9 cells) did indeed lead to the appearance of cation channels (Hu et al. 1994). In keeping with both a role for *trp* and *trpl* in insect phototransduction and with the fact that insect phototransduction is biochemically akin to mammalian signal transduction based on the Gq-PLCβ pathway instead of a transducin-phosphodiesterase (Gt-PDE) pathway (Devary et al. 1987; Selinger and Minke 1988), the *trpl* channels expressed in Sf9 cells could be activated by a Gq-coupled GPCR (Hu and Schilling 1995).

Activation of the Gq-PLCβ signaling system results in hydrolysis of phosphatidylinositol 4,5 bisphosphate (PIP2) with formation of the second messengers diacylglycerol (DAG) and inositol 1,4,5-trisphosphate (IP3), followed by IP3-induced release of Ca^{2+} from intracellular stores and the activation of type-C protein kinases (PKCs) by the combined action of DAG and the released Ca^{2+}. Further, the depletion of intracellular Ca^{2+} stores activates Ca^{2+}-permeable cation channels in the plasma membrane. The molecular basis by which store depletion activates plasma membrane Ca^{2+} entry channels is as yet incompletely defined. Two questions need to be answered: 1) which molecules make up the channels that mediate the store depletion-activated Ca^{2+} entry, and 2) by what mechanism do the stores "inform" the plasma membranes channels of their state of replenishment. TRP channels have been postulated as the pore-forming molecules through which store depletion-activated Ca^{2+} entry takes place (Birnbaumer et al. 1996). While there are data in support of this hypothesis, the final word is not

[1]   Division of Intramural Research, National Institute of Environmental Health Sciences, National Institutes of Health, Department of Health and Human Services, 111 T.W. Alexander Drive, Research Triangle Park, North Carolina 27709

Conn et al.
Insights into Receptor Function
and New Drug Development Targets
© Springer-Verlag Berlin Heidelberg 2006

in, and much is still to be elucidated (italics vide infra the ROCE SOCE Conundrum). On the other hand, recent studies that screened for a Ca^{2+} sensor have been successful in identifying a single pass membrane protein, termed STIM1 (stromal interaction molecule 1), as the store's Ca^{2+} sensor (Liou et al. 2005; Roos et al. 2005). Interestingly, store depletion promotes a translocation of STIM1 from endomembranes to the plasma membrane (Zhang et al. 2005), where it presumably "talks" to the Ca^{2+} entry channels.

Ca^{2+} entering after activation of the PLC-IP3R store depletion pathway serves both as a substrate for the sarcoplasmic-endoplasmic reticulum Ca^{2+} pumps (SERCAs), which replenish the depleted stores, and as a signaling molecule. We will refer to Ca^{2+} entry that follows activation of PLC by receptors as receptor-operated Ca^{2+} entry, or ROCE. Store depletion can also be brought about by inactivating SERCA pumps with an inhibitor such as thapsigargin (Kwan et al. 1990; Thastrup et al. 1990) or by loading cells with Ca^{2+} chelator that acts as a sink for Ca^{2+} that leaks passively from the stores (Hoth and Penner 1992). Both maneuvers activate Ca^{2+} entry, which can be assessed either with a fluorescent indicator dye such as fura2, in which case it is referred to as capacitative Ca^{2+} entry (CCE; Putney 1986, 1990) or as store-operated

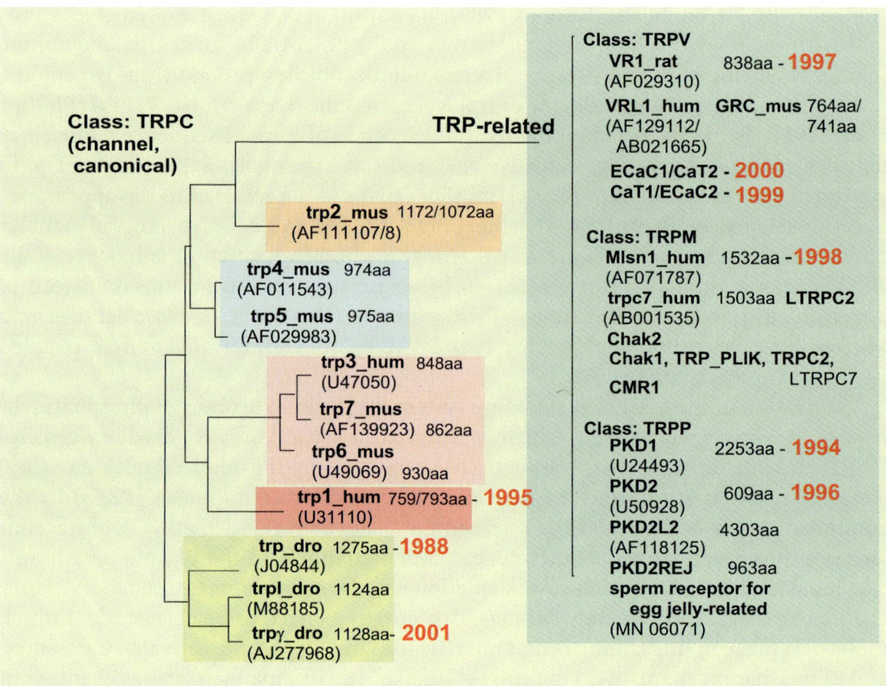

Fig. 1A. **A** Phylogenetic tree illustrating the successive discovery of TRP and the TRP-related channel. Years denote when the founding member of each subfamily was identified. **B** Functional diversity among the major subfamilies of TRP channels. The figure highlights not only the diversity of channels but also the diversity of the regulatory signals impinging on their function. Note in **A** that several TRP-related channels were independently discovered by more than one group, leading to multiple GenBank accession numbers and to confusing names. This confusion was resolved in 2002, as shown with the C, V and M nomenclature in **B** (cf. Montell et al. 2002)

Ca^{2+} entry (SOCE), or by electrophysiological means, where Ca^{2+} entry presents itself as an inward Ca^{2+} current, termed Ca^{2+} release-activated Ca^{2+} current Icrac; (Hoth and Penner 1992; Zweifach and Lewis 1993). Icrac and CCE are commonly accepted as being measures of the same phenomenon.

In 1992, Hardie and Minke showed that the missing sustained phase of the eletroretinograms of *trp* mutant Drosophila eyes has as its underlying basis the absence of a Ca^{2+} conductance. In 1993, they formally raised the question whether the *trp* and *trpl* proteins might be functional homologues of capacitative Ca^{2+} entry channels, and by extension, the pore-forming molecules of Icrac channels (Hardie and Minke 1993) in mammalian cells. The finding that a *trpl/trp* chimera could be activated by store depletion in Sf9 cells (Sinkins et al., 1996) lent strong support to Hardie and Minke's hypothesis.

The mammalian homologues of Drosophila *trp* genes (TRPs) were cloned to test Hardie and Minke's hypothesis. Six such homologues were identified in our initial 1995–96 screen (Zhu et al. 1995, 1996; Wes et al. 1995) and a seventh was discovered three years later (Okada et al. 1999). Initially called TRPs, they are now referred to as TRPCs (see below; Montell et al. 2002). Expression and assembly studies have shown that TRPCs can selectively form heteromeric complexes, such as 1:2, 1:3, 1:5, 4:5, 3:6:7

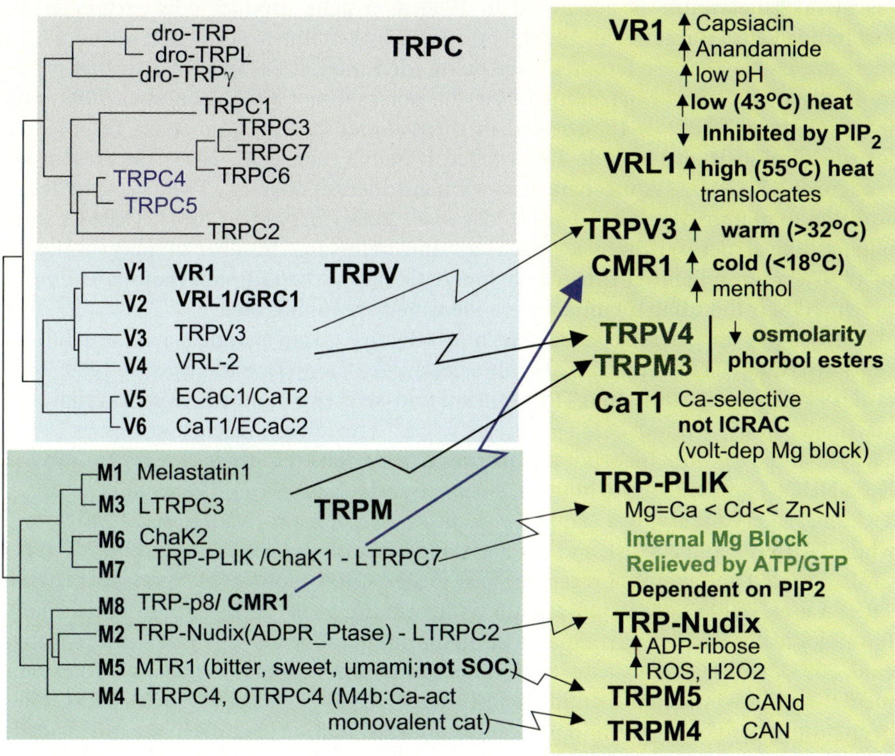

Fig. 1B. (continued)

and 1:4:5, that co-immunoprecipitate (Hofmann et al. 2002; Lintschinger et al. 2000; Struebing et al. 2003). Presumably, in all cases the active channels are tetrameric.

Unexpectedly, independent research from several laboratories uncovered the existence of TRP-related cation channels that together constitute a "superfamily" whose members play differing and sometimes still unknown roles in cellular physiology. One such set of TRP-related channels, with a role in pain and thermo-sensing, was uncovered by researchers in David Julius' laboratory, who in 1997 isolated by expression cloning the cDNA that encodes the capsaicin (also vanilloid) receptor (VR1; Caterina et al. 1997). This finding was followed, shortly afterwards, by the cloning of its close relative, VRL1 (Caterina et al. 1999). Both turned out to be heat sensors and structural homologues of the fly *trp* channels. Independently, VRL1 was also cloned in 1999 as a growth factor-activated channel that translocated from endomembranes to the plasma membrane (Kanzaki et al. 1999). Between 1999 and 2000, two groups, one in The Netherlands and one in Boston, identified two epithelial calcium transporters, renal ECaC (Hoenderop et al. 1999) and intestinal CaT (Peng et al. 1999). Amino acid sequence analysis revealed that ECaC (also CaT2) and CaT (also ECaC2) were related to VR1 and VRL1, and hence to TRPs. In 2001, the original mammalian TRPs were renamed TRPCs (for classic or canonical) and VRs and ECaC/CaT were renamed TRPVs, of which there are six.

Mlsn1 (melastatin 1), identified in 1998 as a gene product down-regulated in melanoma cells (Hunter et al. 1998), is the founding member of the TRPMs, of which there are eight. As these families were being identified, it became evident that other, more distant relatives existed, including the polycystins (Veldhuisen et al. 1999) and their relatives, TRPPs 1–4 (reviewed in Birnbaumer et al. 2003). Genes responsible for mucolipidosis also encode TRP-related proteins (Sun et al. 2000). At present we recognize 4 TRPMLs. A mechanosensory transduction channel, TRPA1, is the latest addition to mammalian TRP genes (Story et al. 2003; Corey et al. 2004). TRPA1's most outstanding structural characteristic is an unusually large number of ankyrin motifs in its N-terminus. Figure 1 shows a phylogenetic tree of the TRP superfamily. The years in which their founding members were identified are highlighted.

Most TRPs are calcium-permeable, non-selective cation channels, with notable exceptions: ECaC/CaT channels are highly selective for Ca^{2+} (Hoenderop et al. 1999; Peng et al. 1999; Yue et al. 2001), and TRPM4/5 are non-selective monovalent cation channels activated by Ca^{2+} without being permeant to Ca^{2+} (reviewed in Fleig and Penner 2004). TRPM2 and TRPM6/7 incorporate enzymatic functions into their C-termini. TRPM2 has a NUDIX domain able to bind ADP–ribose and to sense H_2O_2 (Perraud et al. 2001). TRPM6/7 carry an atypical (alpha) protein kinase domain (Nadler et al. 2001), and while TRPV1, 2 and 3 sense and are activated by distinct temperatures, TRPM8 (also CMR, for cold and menthol receptor) is activated upon cooling (McKemy et al. 2002). TRPV4 and TRPM3 are osmo-sensitive (Strotmann et al. 2000; Grimm et al. 2003). Physiologically, TRPV4 participates in mediating pain sensations (Alessandri-Haber et al. 2003), and TRPM5 is a taste transduction channel expressed in sensory neurons mediating bitter, sweet and amino acid (unami) tastes (Perez et al. 2002). TRPC3 has been proposed to be the melanopsin-activated transduction channel of the intrinsically photosensitive retinal ganglion cells (ipRGCs) that are responsible for entrainment of the circadian clock of the suprachiasmatic nucleus (Qiu et al. 2005; Panda et al. 2005). TRPC3 and TRPC6 were recently identified as essential components of the machinery guiding the Ca^{2+}-dependent growth cone turning of pontine neuron axon extensions (Panda et al. 2005) and Xenopus TRPC1 was identified by Wang and Poo (2005) in a similar phenomenon whereby netrin guides axonal growth of Xenopus spinal neurons. In a parallel study, TRPC3 and TRPC6 were implicated in BDNF-directed axonal outgrowth from cerebellar granule cells (Li et al. 2005). In rodents, TRPC2 has been shown to be the transduction channel activated in vomeronasal sensory neurons in response to activation of vomeronasal pheromone-responsive GPCRs (Liman et al. 1999; Stowers et al. 2002; reviewed in Dulac and Torello 2003 and Zufall et al. 2005). TRPA1, also known as ANKTM1, a channel with 14 N-terminal ankyrin repeats, was characterized as a channel that transduces noxious cold sensation as well as the mechanical bending of stereocilia of inner ear hair cells (Story et al. 2003; Corey et al. 2004). Recently, Kwan et al. analyzed the TRPA1 knockout mice and showed that TRPA1 has a role in mechanical and cold stimuli transduction, but not necessary for hair cell transduction (Kwan et al. 2006).

Mechanism(s) of TRPC activation

Receptor-mediated activation of TRPCs requires phospholipase C activation – A role for PDZ scaffolds?

TRPC channels are activated whenever phospholipase C is activated, either by a Gq-coupled GPCR pathway mediated by PLCβs or by a receptor tyrosine kinase-signaling pathway mediated by the γ family of PLCs. Activation of TRPCs is lost by inhibition of PLCβ with the PLCβ inhibitor U73122 (Zhu et al. 1999) or when activation is tested in systems lacking PLCs, such as the NorpA Drosophila mutant, which lacks PLCβ (Bloomquist et al. 1988), and DT40 chicken B cells, in which PLCγ has been inactivated by gene disruption (Venkatachalam et al. 2003).

In Drosophila, the argument has been made for formation of a "signalplex" with participation of INAD, a multi-PDZ domain-containing scaffold protein, as an intrinsic mechanism by which the *trp/trpl*-based phototransduction channel is activated. In agreement with this postulate, INAD binds to NorpA, PLC, *trp*, *trpl*, rhodopsin and PKC (Shieh and Zhu 1995; Huber et al. 1996; Xu et al. 1998; Chevesich et al. 1997; reviewed in Montell 2001).

In mammals, including man, two of the TRPCs, TRPC4 and TRPC5, also interact with a PDZ scaffold protein, NHERF, the regulator factor of the Na – H exchanger (Tang et al. 2000). NHERF is a two-PDZ domain protein. TRPC4, TRPC5, PLCβ1 and PLCβ2 interact with the first PDZ domain while the other PDZ domain binds members of the Ezrin-Radixin-Moesin (ERM) family of proteins, known to interact with F-actin. Thus, it would appear that, rather than organizing a signalplex similar to INAD's role in the Drosophila eye, NHERF's role in vertebrates may be that of physically connecting members of the PLC-TRP signaling pathway to the cytoskeleton. Yet, this connection is unlikely to be part of the TRPC-activating process, since cortical actin, induced by calyculin A, blocks GPCR as well as store depletion-activated Ca^{2+} entry mediated by TRPC3 (Patterson et al. 1999). Calyculin A is a PP1 and PP2A phosphoprotein phosphatase inhibitor that causes accumulation of the C-terminally phosphorylated forms of ERM proteins. These in turn interact with F-actin, promoting its redistribution to the plasma membrane where the N-terminal portions of ERM proteins interact with membrane proteins. As a consequence, calyculin A treatment leads to the formation of cortical actin. The implications of establishing independent connections between either PLCs and the actin cytoskeleton or TRPC4-5 and the actin skeleton, which could interfere with TRPC's function as an ion channel, are not clear.

However, the formation of transient complexes is likely to be involved in the process by which TRPCs are activated. An analysis of TRPC6 activation by Gq-coupled M1 muscarinic receptors in PC12D cells by Kim and Saffen(2005) showed the formation of time-sensitive macromolecular complexes involving PKC and phosphorylation of TRPC6 at a conserved PSPK site 23 amino acids downstream of the EWFKAR motif. These authors found that, once PKC was dissociated from the phosphorylated channel, the $PS(PO_3^-)PK$ motif of TRPC6 recruited the FK506-binding protein FKBP12, which in turn recruited calcineurin (CN) and calmodulin (CaM). CN then dephosphorylated TRPC6, causing dissociation of the complex into its individual components. The complexes did not form if the Ser of the PSPK motif was mutated, when PKC was inhibited, when the immunophilin FKBP12 was blocked with FK506 or rapamycin, or when cells had been treated with cyclosporin. In all these instances, the channel was not dephosphorylated and the M1R stayed associated with the channel.

Activation by conformational coupling – role for IP3 receptor as an activator of TRPC through protein–protein interaction

The conformational coupling concept emerged originally from the elucidation of the role of the skeletal muscle voltage-gated Ca^{2+} channel in mediating a depolarization-induced contraction. Skeletal muscle fibers have a particularly strong Ca^{2+} uptake activity performed by SERCA pumps. As a consequence, the Ca^{2+} released from the sarcoplasmic reticulum (SR) in response to Ca^{2+}-activated Ca^{2+} release through the ryanodine receptor (RYR)/Ca^{2+} release channel is almost quantitatively reabsorbed into the SR with little or no loss through the action of plasma membrane Ca^{2+} pumps. As a consequence, isolated skeletal muscle fibers contract repeatedly in response to repeated depolarizing stimuli. Initially, the Ca^{2+} that triggers Ca^{2+}-activated Ca^{2+} release through the RYR was thought to come from the extracellular milieu, admitted through the skeletal muscle voltage-gated Ca^{2+} channel (CaV1.1). In support of this theory, depolarizing in the presence of a CaV1.1 channel blocker, such as a dihydropyridine (DHP), abolished the response to depolarization. It was thus surprising that contractions that were blocked by DHPs could still be elicited in totally Ca^{2+} free media. Activation of the RYR was thus not dependent on an initial Ca^{2+} influx through the channel, but was nevertheless dependent on the presence of a fully functional voltage- and DHP-sensitive complex. The molecular basis for this phenomenon became clear when it was discovered that, upon membrane depolarization, the Ca^{2+} channel changed its conformation and interacted physically with the nearby RYR, triggering its initial opening, initial Ca^{2+} release and the ensuing explosive Ca^{2+}-activated Ca^{2+} release responsible for activation of the actomyosin contractile machinery (reviewed in Rios and Pizarro 1991).

The conformational coupling model for TRPC activation is based on the skeletal muscle excitation-coupling model, in which the information flow is from a plasma membrane ion channel (CaV1.1) to an endomembrane Ca^{2+} release channel (RYR), but with the flow of information, i.e., the sense of the signaling pathway, inverted. The model postulates that the same stimulus that activates the IP3R to release Ca^{2+} from the endoplasmic reticulum Ca^{2+} stores, IP3, also activates a TRPC-activating function of the IP3 receptor. Activation of TRPCs would come about by physical binding of a TRPC-binding domain of the IP3R to TRPC with attendant activation of the affected TRPC.

This hypothesis was tested in our laboratory in the late 1990s using a GST pull-down approach (Boulay et al. 1999). We found that a region on the post-transmembrane C-terminal domain of TRPCs interacts with a region located between (IP3R-3 numbering) amino acids 675 and 800 of the IP3Rs. This region lies circa 100 amino acids C-terminal to the IP3 binding domain (amino acids 225–575) distal to the C-terminally located ion channel forming transmembrane domains (amino acids 224–2,565) of the 2,761 amino acid IP3R-2. More importantly, transient overexpression of GST-fusion fragments of TRPC-interacting sequences of the IP3R either inhibited or extended the duration of Ca^{2+} influx through endogenous receptor- or store depletion-activated Ca^{2+} entry channels (Boulay et al. 1999). Thus, IP3R sequences identified as TRPC-binding sequences affect Ca^{2+} entry, which is postulated to be mediated by TRPC channels. The regions identified on IP3R-2 are conserved in IP3R-1 and IP3R-3, and the sequences identified on TRPC3 as interacting with IP3Rs are preserved in other TRPCs.

Sage's laboratory (Rosado and Sage 2001) used a different approach to test the conformational coupling hypothesis. Working with human platelets, they probed for co-immunoprecipitation of endogenous TRPC1 with endogenous IP3R. No IP3R co-immunoprecipitated from control lysates of control platelets, but activation of Ca^{2+} influx secondary to store depletion, induced by inactivation of SERCA pumps with thapsigargin, resulted in IP3R co-immunoprecipitating with the human TRPC1. As was the case in earlier experiments with A7r5 and DDT1-MF2 smooth muscle cell lines (Patterson et al. 1999), induction of cortical actin in platelets with jasplakinolide, blocked the interaction of the IP3R with TRPC1. These experiments are especially relevant because the activation-dependent interaction of IP3R with TRPC was shown in a normal cell with normal complements of the interacting partners. Therefore, interactions observed under these conditions do not suffer from the drawbacks of interactions shown only upon overexpression of the interacting partners. It remains to be determined how IP3R-mediated activation of TRPC channels interfaces with STIM1 translocation and activation of Icrac.

Activation of TRPCs by diacylglycerol and inhibition of TRPCs by protein kinase C

In 1999, Hofmann et al. (1999) reported that diacylglycerols (DAG), one of the two reaction products resulting from PLC activation, activate TRPC3 and TRPC6 and do so independent of protein kinase (PKC) activation. That same year, Okada et al. (1999) reported the same finding for the newly cloned TRPC7. In both studies, the effects of DAGs could be augmented by inhibition of DAG lipase or DAG kinase, thus sparing DAG removal, and were insensitive to PKC inhibitors.

The study by Okada et al. (1999) also showed that activation of PKC inhibited activation of TRPC7 by subsequently added DAG, indicating opposite roles for PKC and DAG.

A superficial examination of DAG's effects on TRPC4 and TRPC5 does not show stimulation by DAG. Yet upon inhibition of PKCs, both TRPCs are robustly activated by DAG (Venkatachalam et al. 2003; Okada et al. 1999). Indeed, activation of PKC inactivates all TRPC's so far studied for this effect (TRPC3 through TRPC7), not only for activation by DAG but also by the Gq-coupled and the receptor tyrosine kinase (RTK) signaling pathways (Okada et al. 1999; Venkatachalam et al. 2003; Trebak et al. 2005). But, the rates at which DAG and PKC act on the TRPCs differ with TRPC subtype. Thus, for the TRPC3 family of TRPCs (TRPC3, TRPC6 and TRPC7), activation by DAG is faster than phopshorylation by PKC, so activation by DAG is the dominant phenomenon. On the other hand, for TRPC4 and TRPC5, the DAG-induced inhibition by PKC is established before DAG can activate the TRPC channel. Whether this differs after induction of cortical actin has not been reported.

DAG also activates the TRPC2 channel in its native environment, the vomeronasal sensory neuron (Lucas et al. 2003). The effect is fast, potentiated by DAG lipase and DAG kinase inhibitors and unaffected by inhibitors of PKC. Whether PKC is inhibitory to TRPC2, as it is for TRPC3 through TRPC7, has not been reported. Given that the Ser phosphorylated by PKC is located in a motif that is conserved in all TRPCs (Trebak et al. 2005), it is tempting to suggest that PKC is inhibitory to all. In agreement with the interpretation that the channel being activated by DAG in vomeronasal sensory neurons is indeed TRPC2, genetic ablation of TRPC2 resulted in loss of the DAG activated current (Lucas et al. 2003).

Activation of Ca^{2+} entry by channel translocation from endomembranes to the plasma membrane

Insulin stimulates glucose uptake into its target tissues by promoting the translocation of GLUT4-bearing endovesicles to the plasma membrane. The incorporation of AMPA-type glutamate receptors into the postsynaptic membrane of neurons undergoing high frequency stimulation is the basis for the establishment of early long-term memory (eLTP) in hippocampal CA1 neurons. The TRPV2 channel, also called growth factor regulated channel or GRC, transitions from internal membranes to the plasma membrane upon treatment of myotubes with insulin-like growth factor-1 (IGF-1; Kanzaki et al. 1999). Boulay and coworkers (Cayouette et al. 2004) tested whether TRPC6 might be under similar control and indeed saw an increase of cell surface TRPC6, as seen by an increase of biotinylated TRPC6 by immunostaining within 30 sec of stimulating cells either with an agonist for a Gq-coupled GPCR (M5R) or with the store-depleting agent thapsigargin. In a different context, Clapham and collaborators showed that activation of Rac1 in hippocampal neurons leads to translocation of TRPC5 from endomembranes to the plasma membrane in a process that involves activation by Rac1 of PIP(5)Kα (phophatidylinositol-4-phosphate 5-kinase α) and synthesis of PIP2 (phosphatidylintositol-4,5-bisphosphate). Rac1 in turn is activated by stimulation of an RTK (e.g., EGFR) activating PI3K with formation of PIP3 (phosphatidyl-inositol 3,4,5 tris-phosphate), which in turn activates a Rac1 guanine nucleotide exchange factor (Rac-GEF), leading to augmented GTP-Rac1 (Bezzerides et al. 2004). This signalling mechanism (RTK to TRPC5 incorporation into plasma membrane) has been implicated in Ca^{2+}-dependent repression of neurite outgrowth, as there is an inverse correlation of PIP(5)Kα with neurite length.

Physiological role for TRPC channels as electrogenic devices that couple GPCR-Gq activation to voltage-gated Ca^2+ channel activation

Activation of non-selective monovalent cation channels leads to a collapse of the membrane potential. Many TRP and TRP-related channels are mostly non-selective cation channels, some with selectivity for monovalent cations over divalents, some non-selective with respect to both monovalent and divalent cations. As such, activation of these types of channels has in common that it dissipates the transmembrane potential of the cells in which the channels are expressed. This property was highlighted in a report from the Fleig-Penner laboratory, in which the channel properties of TRPM5, a Ca^{2+}-activated non-selective monovalent cation channel, i.e., a CAN, was characterized (Launay et al. 2002). More recently, Soboloff et al. (2005), studying the effect of down regulating TRPC6 with small interfering RNA (siRNA), found that, in A7r5 smooth muscle cells, TRPC6 fulfills the role of an electrogenic coupling mechanism coupling a Gq-coupled GPCR to Ca^{2+} influx through a dihydropyridne-sensitive Ca^{2+} channel. This role became apparent when, upon siRNA treatment, they saw a > 90% reduction of muscarinic receptor-stimulated TRPC6 channel activity, as measured by the patch clamp technique, with essentially no loss of Ca^{2+} influx, as measured with the fluorescent Ca^{2+} indicator dye, fura2. Muscarinic receptor-stimulated Ca^{2+} influx into siRNA-treated cells was completely inhibited by a dihydropyridine Ca^{2+} channel blocker (see Soboloff et al. 2005). While this is the first demonstration of an electrogenic coupling role for a TRPC channel, it is likely that other examples are soon to follow, especially in natural tissue cells.

A Role for Tyrosine Kinases in Voltage-Gated Ca^{2+} channel-independent Ca^{2+} influx

The original observation that tyrosine phosphorylation may be an important regulator participating in activation of receptor- and store-operated Ca^{2+} entries (ROCE and SOCE) in non-excitable cells came from studies showing an inhibitory effect of tyrosine kinase inhibitors on these forms of calcium entry in human foreskin fibroblasts (Lee et al. 1993; Lee and Villereal 1996). This observation was followed by studies that showed total loss of bradykinin-stimulated ROCE and partial absence of SOCE in embryonic fibroblasts derived from mice lacking the *src* tyrosine kinase (Babnigg et al. 1997). Given that evidence has accumulated in the literature showing that tyrosine phosphorylation is a common consequence associated with stimulation of cells via receptors that signal by using the Gq-PLC-calcium mobilizing pathway (Gutkind and Robbins 1992; Igishi and Gutkind 1998), we became interested in the possibility that tyrosine phosphorylation may be an activating signal for one of the events that leads to receptor- and/or store-operated calcium entry. Indeed, experiments parallel to ours, testing for a role of tyrosine kinases in the receptor-mediated activation of the type 3 TRPC (TRPC3), revealed that in HEK cells, the activation of this TRPC by a PLC-stimulating GPCR is inhibited by inhibitors of tyrosine kinases, and that, when expressed in *src* kinase-negative cells, the transfected TRPC3 is not activated by a co-transfected Gq-coupled receptor (Vazquez et al. 2004). This observation recapitulated the earlier findings with an endogenous (bradykinin-activated) GPCR acting via Gq activation on the endogenous complement of the receptor-operated Ca^{2+} entry pathway (Babnigg et al. 1997).

Studies on the Role(s) of Tyrosine Phosphorylation in TRPC Function

To learn more about the possible role of tyrosine kinase(s) in TRPC-mediated events, as well as in SOCE, in the last few years we have used *in vitro* and in cell protein-protein interaction assays (Kawasaki et al. 2006). We found that *c-src* phosphorylates TRPC3 on tyrosine 226 (Y226) located on the TRPC3 N-terminus and that formation of phospho-Y226 is essential for TRPC3 activation. Surprisingly, this requirement is unique for TRPC3, since 1) mutation of the cognate tyrosines of the closely related TRPC6 and TRPC7 channels had no effect on their function, 2) both TRPC6 and TRPC7 were activated in *src*-, *yes*- and *fyn*-deficient cells (SYF cells), and 3) *src*, but not *yes* or *fyn* (YF cells), rescued TRPC3 activation in cells lacking *src*, *yes* and *fyn*. Yet, we found that the SH2 domain of *c-src* interacted not only with TRPC3 but also with either the N-termini or the C-termini of all other TRPCs. This finding suggests that other tyrosine kinases may play a role in ion fluxes mediated by TRPCs.

A side-by-side comparison of the effects of genistein on endogenous ROCE and SOCE in YF mouse embryonic fibroblast (MEF), HEK and COS-7 cells showed these influxes to be inhibitable in all three cells types but with differing sensitivities (Figs. 2 and 3). Taken together these results argue for the channels mediating ROCE and SOCE to be heterogeneous and to differ from tissue to tissue.

The finding that TRPC6 is active in SYF and YF cells was unexpected, as it has been shown to be a substrate of *fyn* (Hisatsune et al. 2004) and to behave essentially the same as TRPC3 in in vitro and cell expression assays. Thus, TRPC6 is phosphorylated by co-expression with *fyn* in COS cells and associates with *fyn* in GST pulldown assays by interaction of its N-terminus with the SH2 domain of *fyn* (Hisatsune et al. 2004), and TRPC6 activation is inhibited by the tyrosine kinase inhibitor PP2, regardless of whether it is activated by a receptor-tyrosine kinase-PLCγ pathway (triggered by EGF; Hisatsune et al. 2004) or by the DAG pathway (Soboloff et al. 2005). Further, addition

---

Fig. 2. Comparison of the inhibitory effect of genistein on ROCE and SOCE endogenous to HEK cells, COS-7 cells and mouse embryonic fibroblasts (MEFs) lacking the indicated members of the *src*-family of tyrosine kinases. In the experiments shown in this and the other figures, ROCE was assessed in cells expressing the indicated Gq-coupled GPCR in transient or stable form, loading the cells that had been plated on coverslips with fura2 and subjecting these cells to a Ca^{2+} mobilization protocol in which the PLCβ system was activated by the cognate receptor agonist (carbachol (CCh) for the M5 muscarinic recepetor and arginine vasopressin (AVP) for the V1a vasopressin receptor) in the absence of external Ca^{2+}. $[Ca^{2+}]_i$ changes were then followed by video spectromicroscopy to record Ca^{2+} release from internal stores; $[Ca^{2+}]$ was allowed to return to near basal levels. At this point, Ca^{2+} was added to the external medium, and the influx of Ca^{2+}, leading to increase in $[Ca^{2+}]_i$ representing ROCE, was monitored for the indicated times. When drugs such as genistein or KB-R7943 (see below) were added, they were present throughout the first and second phases of the $[Ca^{2+}]_i$ changes. Note that only the Ca^{2+} entry phases are shown. In none of the experiments shown in this or the other figures did the presence of tyrosine kinase or Na-Ca exchange inhibitors significantly affect the IP3- or thapsigargin-induced Ca^{2+} release from the endogenous stores. SOCE was assessed by substituting the GPCR agonist for thapsigargin. Gd^{3+} (5–10 μM) was added when TRPC-mediated ROCE was measured. At this concentration of Gd^{3+}, TRPC3, 5, 6 and 7 mediated Ca^{2+} entry is unaffected, but endogenous ROCE is inhibited. Data on effects of genistein are either unpublished or adapted from Kawasaki et al. (2006)

of *fyn* to inside-out membrane patches from cells expressing TRPC6 increased basal and oleyl-acetyl-glyceride (OAG) stimulated TRPC6 activity (Hisatsune et al. 2004). Yet, TRPC6-ROCE is activated in cells lacking not only *fyn* but also *yes* and *src* (Figs. 2 and 3). It is currently not known whether this discrepancy is due to our use of the GPCR-Gq-PLCβ activation pathway instead of a RTK-PLCγ pathway (which might be impaired in SYF cells) or is due to the different form of assessing TRPC6 ROCE, fura2 in our case and electrophysiological in inside-out membrane patches in the case of Hisatsune et al. (2004). If, however, the lack of sensitivity of TRPC6 to genistein in our experiments,

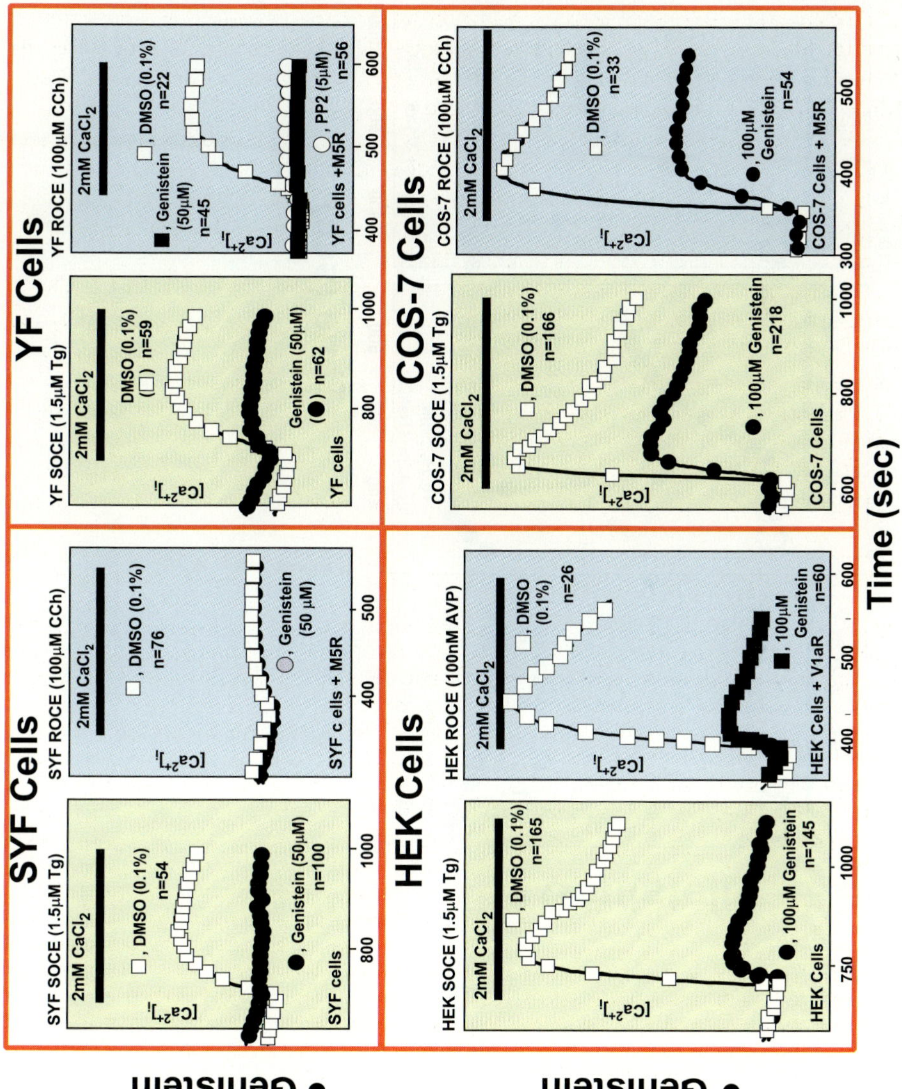

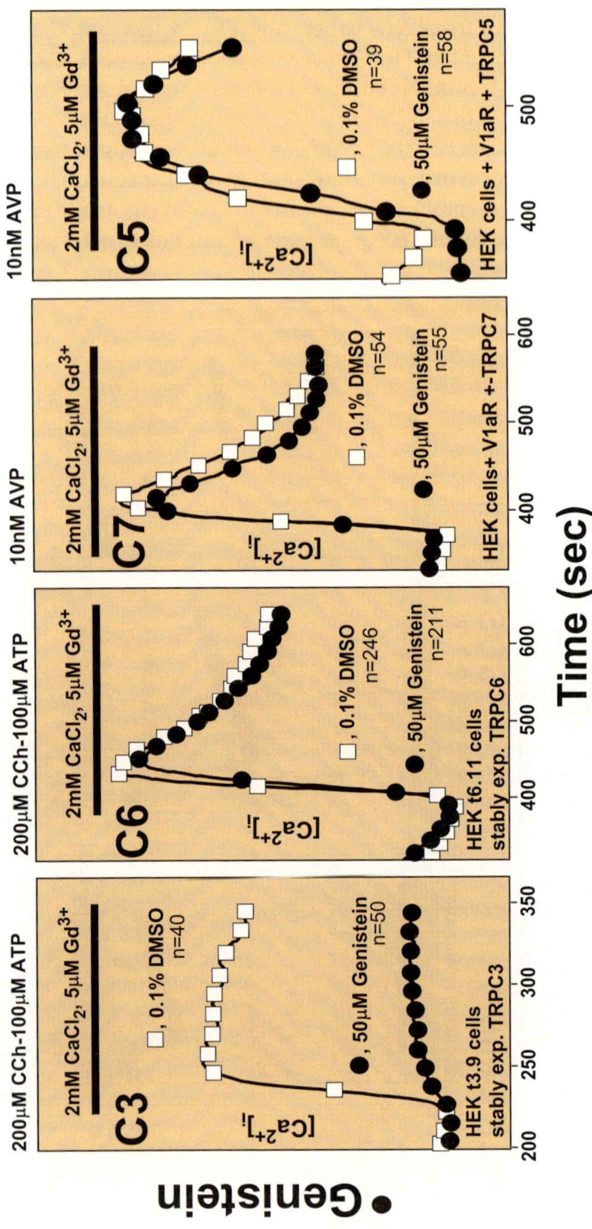

Fig. 3. Comparison of the inhibitory effect, or lack thereof, of genistein on ROCE mediated by the indicated TRPC channels, as seen in HEK cells. (Adapted from Kawasaki et al. 2006)

and TRPC6 being active in SYF cells is not due to the use of differing activation pathways or the method used to assess TRPC6 activation, the data may also indicate that functionally, in addition to *src*, *fyn* and *yes*, there is at least one more "PP2-sensitive *src*-family tyrosine kinase" able to regulate TRPC6 and other TRPCs. As mentioned, the fact that TRPC1 through C7 all interacted with *src* in GST pulldown assays raises the possibility that all TRPC channels depend on tyrosine phosphorylation for their functioning as an effector system for the activation of the GPCR-Gq-PLCβ/RTK-PLCγ pathways.

TRPCs and SOCE: Concluding Remarks

Table 1 summarizes properties of TRPC channels in terms of which signalling pathways have been shown to activate each channel and possible mechanisms by which they are activated. Two pathways feed into TRPCs. One generates DAG from the action of PLCβ activated by Gq and Gi-derived Gβγ and the other generates DAG from the action of PLCγ activated by tyrosine phosphorylation either directly by the RTK-type receptor or, secondarily, to non-RTK tyrosine kinase recruitment such as happens upon T- and B-cell receptor activation. Activation of TRPCs by DAG may be aided by the action of IP3R, acting by protein:protein interaction according to a conformational coupling model, and by the formation of a multimolecular signalling complex with or without involvement of a protein acting as nucleating scaffold.

A direct IP3R:TRPC interaction, especially if facilitated by the action of STIM1 (Liou et al. 2005; Zhang et al. 2005; Roos et al. 2005), may also account for activation of TRPC channels by store depletion.

In contrast to the rather satisfactory models one can set up for explaining how TRPC channels may be activated, there are only scant data suggesting how, if at all, TRPC channels participate in SOCE. The strongest data set was recently published by Villereal's laboratory studying the natural channels that contribute to SOCE in HEK 293 cells, testing for interference with siRNA. More that > 80% down-regulation of the naturally expressed TRPC1, TRPC3 and TRPC7 proteins resulted in a circa > 50% reduction in thapsigargin-induced SOCE (Zagranichnaya et al. 2005). The > 50% loss is curious, as it happens to coincide with the similar loss of acetyl choline-induced, NO-mediated vascular relaxation of aortic rings seen in mice lacking TRPC4 (Freichel et al. 2001).

The missing information or hypothesis relating TRPCs to SOCE is how non-selective cation channels come together to form a highly Ca^{2+}-selective ion channel. On a purely speculative level, there are three possibilities that come to mind. One is that ion selectivity does change in channels that are heteromeric in nature. The other is that ion selectivity is altered by post-translational modification. Dietrich et al. (2003) have shown that basal or constitutive activity is affected by glycosylation. Kawasaki et al. (2006) found that at least one channel, TRPC3, is dependent on tyrosine phosphorylation for activity. The effect of compound kinase actions has not been explored. Groschner and collaborators, by showing that in HEK cells TRPC3-mediated Ca^{2+} influx is dependent on extracellular Ca^{2+} and a functional Na–Ca exchanger operating in reverse (Rosker et al. 2004), raised the possibility that the Ca^{2+} selectivity is the result of a tandem arrangement whereby Na^+ entering through a TRPC channel is

Table 1. Regulation of TRPC channels. Comparison to endogenous ROCE and SOCE of HEK cells, *yes*- and *fyn*-deficient mouse embryonic fibroblasts (YF MEFs), COS-7 cells and HSWP human foreskin fibroblasts

Channel	Activation by store depletion[b]	Stimulation by Gq-coupled GPCR[c]	2nd messenger	Effect of PKC	Activation by RTK[a]	Inhibition by genistein PP2	Highly Ca^{2+}-selective?	Channel block by Gd^{3+} (5–10µM)	Conformat'l coupling (interaction with IP3Rs)
TRPC1	yes[1]	yes	?	?	yes	?	No	Inh.?	yes
TRPC2	yes[2,3]	yes	DAG[9]	?	yes	?	No	Inh.?	yes
TRPC3	yes[d4]	yes	DAG[10]	Inh.[11]	yes	yes-Y226[13]	No	Resistant	yes
TRPC3a	yes***[5]	yes	nt	nt	yes	nt	No	Resistant	yes
TRPC4	yes[6]	yes	DAG[11]	Inh.[11]	yes	?	No	? Activated @100µM[14]	yes
TRPC5	yes[7]	yes	DAG[11]	Inh.[11]	yes	No[13]	No	Resistant Activated @100µM[15]	yes
TRPC6	?	yes	DAG[10]	Inh.[12]	yes	No[13], yes[12]	No	Resistant[13], Inh.[16]	yes
TRPC7	yes[d8]	yes	DAG[8]	Inh.[8]	yes	No[13]	No	Resistant	yes
ROCE (Gq-coupled Rs): HEK,YF MEF,COS,HSWP[17]						Inhibition	No	Inhibition	
SOCE (thapsigargin): HEK,YF MEF,COS,HSWP						Inhibition	Yes	Inhibition	

[a] EGF receptor (EGF), B cell receptor (anti-IgM), T cell receptor (anti-CD3).

[b] thapsigargin;

[c] M5 muscarinic or V1a vasopressin receptors;

[d] requires low level of expression.

1, Liu et al. 2000; Rosado and Sage 2001; 2, Vannier et al. 1999; 3, Jungnickel et al. 2001; 4, Vazquez et al. 2003; 5, Yildirim et al. 2005; 6, Philipp et al. 1996; 7, Philipp et al. 1998; 8, Okada et al. 1999; Riccio et al. 2002b; Lievremont et al. 2004; 9, Lucas et al. 2003; 10, Hofmann et al. 1999; 11, Venkatachalam et al. 2005; 12, Soboloff et al. 2005; 13. Kawasaki et al. 2006; 14, Schaefer et al. 2002; 15, Jung et al. 2003; 16, Inoue et al. 2001; 17, Lee et al. 1993. Actv'n, activation; Inh., inhibition; Stimul'n, stimulation; Conformat'l, conformational; nt, not tested. HSWP, human foreskin fibroblasts.

extruded not only by the Na–K ATPase but also by the Na–Ca exchanger, an exquisitely Ca^{2+}-selective machine.

The third possibility regarding the molecular makeup of ROCE and, especially, SOCE channels is that researchers may have been somewhat naive in assuming that there is only one Icrac channel. The fact that inhibition of endogenous SOCE by genistein shows varying degrees of sensitivity may be an indication that SOCE channels – presumed to be equivalent to Icrac channels – are heterogeneous in nature. If so,

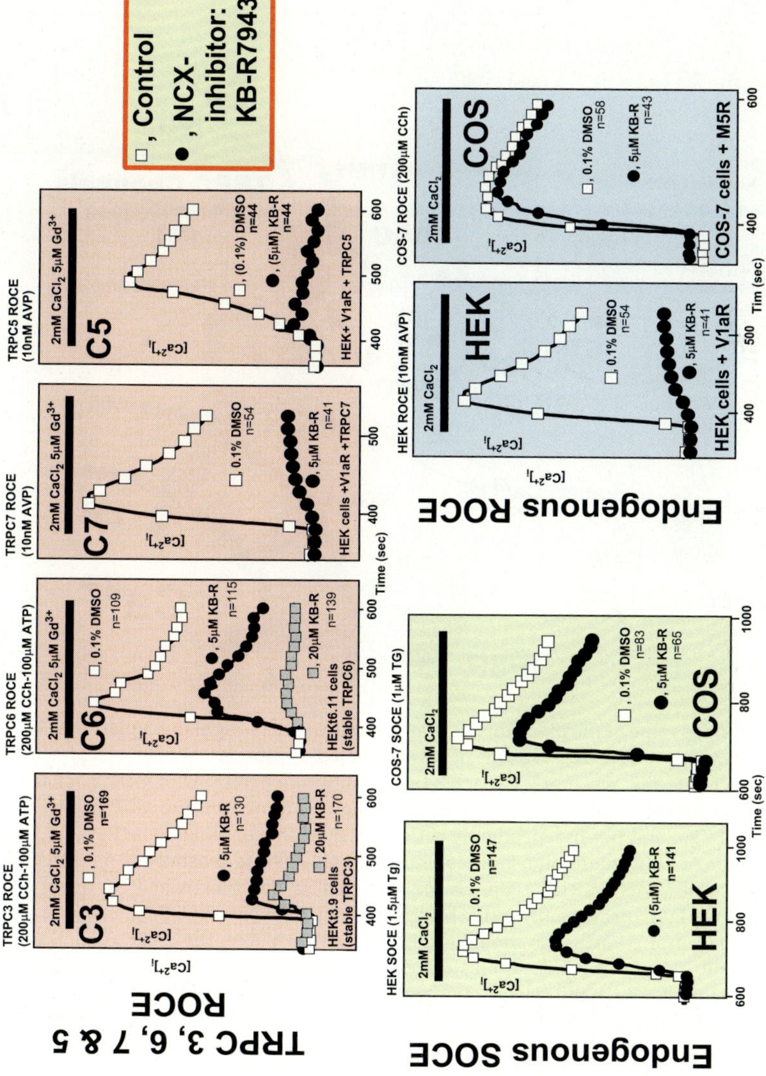

Fig. 4. Comparison of the effect of the partially selective Na–Ca exchange inhibitor, KB–R7943, on ROCE mediated by TRPCs and on ROCE and SOCE mediated by endogenous channels

TRP-related channels other than TRPCs may form SOCE channels. The participation of various TRPVs, especially TRPV5 and TRPV6, comes to mind. In line with this reasoning, Schindl et al. (2002) noted commonalities between TRPV6 (CaT1) and Icrac. The studies reported by Rosker et al. (2004) implicating the Na-Ca exchanger in ROCE mediated by TRPC3 in HEK cells prompted us to test the effectiveness with which the Na–Ca exchanger (NCX) inhibitor, KB-R7943, affects ROCEs mediated by several TRPCs in HEK cells and ROCEs mediated by endogenous ROCE channels in HEK and other cells. The picture that emerged is not one that supports an obligatory role for Na–Ca exchangers in ROCE or SOCE – for this KB-R7943 is too non-specific – but one

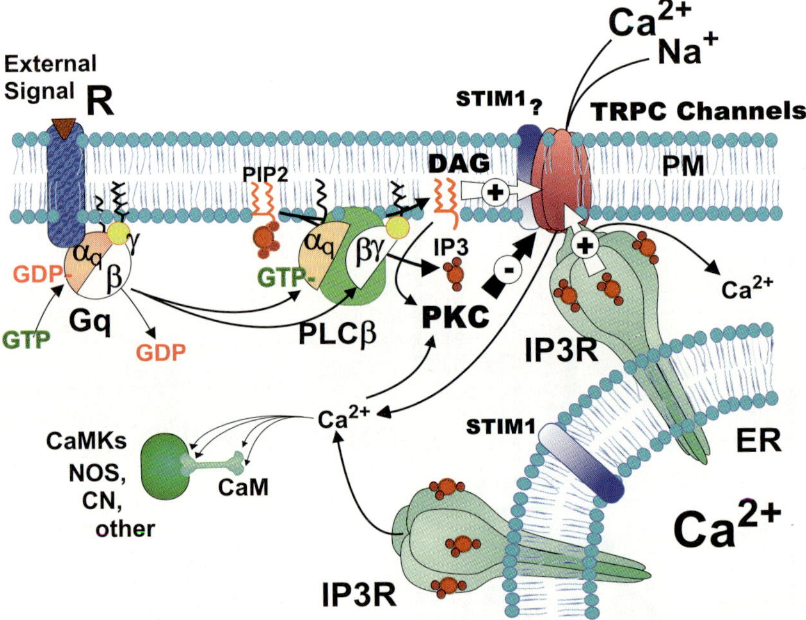

Fig. 5. Overview of some of the mechanisms that regulate or are thought to regulate the Gq-PLCβ triggered activation of TRPCs. A ligand of a Gq-coupled G-protein coupled receptor (GPCR) is shown to activate the receptor's Gq-activating function, whereby Gq's GDP is changed to GTP with concomitant dissociation of the heterotrimeric Gq protein into GTP-α plus the β-γ dimer. Both GTP-αq and β-γ independently activate β-type phospholipase Cs (PLC βs), leading to the hydrolysis of phosphatidylinositol bis-phosphate (PIP2) into inositol-trisphosphate (IP3) plus diacylglycerol (DAG). TRPC channels are depicted as being activated by three distinct mechanisms: 1) by DAG, presumably acting by direct interaction with the TRPC, 2) by the Inositol trisphosphate receptor (IP3R), also thought to interact directly with the TRPC, and 3) possibly by STIM1, an ER Ca sensor, which upon store depletion is translocated from the endoplasmic reticulum (ER) membrane to the plasma membrane (PM). The figure also highlights the negative regulation by PKC, activated by the cooperative interaction effect of DAG (generated by the action of the PLCβ) and Ca^{2+}, originating first from the store and later from the extracellular milieu entering through the TRPC. CAMKs, Ca-calmodulin activated kinases; NOS, nitric oxide synthase; CN, calcineurin – also PP2B

that, once more, suggests that endogenous channels mediating ROCE and SOCE are likely to be heterogeneous in their molecular makeup (Fig. 4).

ROCE and SOCE are forms of Ca^{2+} entry with specific functions in a large list of diverse cellular functions that include smooth muscle contraction, B- and T-cell activation, and vascular permeability and development of the central nervous system, to name a few. Yet, the molecular make-up of the channels mediating ROCE and SOCE is still largely a matter of speculation. New tools are needed to confirm or deny the hypothesis that TRPCs alone or in combination with other members of the TRP superfamily participate in these forms of Ca^{2+} entry. In the context of this IPSEN Conference on "Insights into Receptor Function and New Drug Development Targets," we see TRPCs as long-overdue targets for pharmacological intervention. Finally, Fig. 5 summarizes the known mechanisms for regulating TRPC channel function.

Acknowledgements. Supported by the Intramural Research Program of the NIH and by NIEHS.

References

Alessandri-Haber N, Yeh JJ, Boyd AE, Parada CA, Chen X, Reichling DB, Levine JD (2003) Hypotonicity induces TRPV4-mediated nociception in rat. Neuron 39:497–511

Babnigg G, Bowersox SR, Villereal ML (1997) Role of pp60^{c-src} in the regulation of calcium via store-operated calcium channles. J Biol Chem 272:29434–29437

Bezzerides V, Ramsey S, Greka A, Clapham DE (2004) Rapid translocation and insertion of TRPC5 channels. Nature Cell Biol 6:709–720

Birnbaumer L, Zhu X, Jiang M, Boulay G, Peyton M, Vannier B, Brown D, Platano D, Sadeghi H, Stefani E, Birnbaumer M (1996) On the molecular basis and regulation of cellular capacitative calcium entry: roles for Trp proteins. Proc Natl Acad Sci USA93:15195–15202

Birnbaumer L, Yildirim E, Abramowitz J (2003) A comparison of the genes coding for canonical TRP channels and their M, V and D relatives. Cell Calcium 33:419–432

Bloomquist BT, Shortridge RD, Schneuwly S, Perdew M, Montell C, Steller H, Rubin G, Pak WL (1988) Isolation of a putative phospholipase C gene of Drosophila, norpA, and its role in phototransduction. Cell 54:723–733

Boulay G, Brown DM, Qin N, Jiang M, Dietrich A, Zhu MX, Chen Z, Birnbaumer M, Mikoshiba K, Birnbaumer L (1999) Modulation of Ca^{2+} entry by polypeptides of the inositol 1,4,5-trisphosphate receptor (IP3R) that bind transient receptor potential (TRP): evidence for roles of TRP and IP3R in store depletion-activated Ca^{2+} entry. Proc Natl Acad Sci USA 96:14955–14960

Caterina MJ, Schumacher MA, Tominaga M, Rosen TA, Levine JD, Julius D (1997) The capsaicin receptor: a heat-activated ion channel in the pain pathway. Nature 389:816–824

Caterina MJ, Rosen TA, Tominaga M, Brake AJ, Julius D (1999) A capsaicin-receptor homologue with a high threshold for noxious heat. Nature 398:436–441

Cayouette S, Lussier MP, Mathieu EL, Bousquet SM, Boulay G (2004) Exocytotic insertion of TRPC6 channel into the plasma membrane upon Gq protein-coupled receptor activation. J Biol Chem 279:7241–7246

Chevesich J, Kreuz AJ, Montell C (1997) Requirement for the PDZ domain protein, INAD, for localization of the TRP store-operated channel to a signaling complex. Neuron 18:95–105

Corey DP, Garcia-Anoveros J, Holt JR, Kwan KY, Lin S-Y, Vollrath MA, Amalfitano A, Cheung EL-M, Derfler1 BH, Duggan A, Geleoc GSG, Gray PA, Hoffman MP, Rehm HL, Tamasauskas D, Zhang DS (2004) TRPA1 is a candidate for the mechanosensitive transduction channel of vertebrate hair cells. Nature 342:723–730

Devary O, Heichal O, Blumenfeld A, Cassel D, Suss E, Barash S, Rubinstein CT, Minke B, Selinger Z (1987) Coupling of photoexcited rhodopsin to inositol phospholipid hydrolysis in fly photoreceptors. Proc Natl Acad Sci USA 84:6939–6943

Dietrich A, Mederos y Schnitzler M, Emmel J, Kalwa H, Hofmann T, Gudermann T (2003) N-Linked protein glycosylation is a major determinant for basal TRPC3 and TRPC6 channel activity. J Biol Chem 278:47842–47852

Dulac C, Torello AT (2003) Molecular detection of pheromone signals in mammals: from genes to behavior. Nature Rev Neurosci 4:551–562

Fleig A, Penner R (2004) The TRPM ion channel subfamily: molecular, biophysical and functional features. Trends Pharmacol Sci 25:633–639

Freichel M, Vennekens R, Olausson J, Hoffmann M, Muller C, Stolz S, Scheunemann J, Weissgerber P, Flockerzi V (2004) Functional role of TRPC proteins in vivo: lessons from TRPC4-deficient mouse models. Biochem Biophys Res Commun 322:1352–1358

Grimm C, Kraft R, Sauerbruch S, Schultz G, Harteneck C (2003) Molecular and functional characterization of the melastatin-related cation channel TRPM3. J Biol Chem 278:21493–21501

Gutkind JS, Robbins KC (1992) Activation of transforming G-protein coupled receptors induces rapid tyrosine phosphorylation of cellular proteins including p125^{FAK} and p130^{src} substrate. Biochem Biophys Res Commun 188:155–161

Hardie RC, Minke B (1992) The *trp* gene is essential for a light activated Ca^{2+}-channel in Drosophila photoreceptor cells. Neuron 8:643–651

Hardie RC, Minke B (1993) Novel Ca^{2+} channels underlying transduction in *Drosophila* photoreceptors: implications for Phosphoinositide-mediated Ca^{2+} mobilization. Trends Neurosci 9:371–376

Hisatsune C, Kuroda Y, Nakamura K, Inoue T, Nakamura T, Michikawa T, Mitsutani A, Mikoshiba K (2004) Regulation of TRPC6 channel activity by tyrosine phosphorylation. J Biol Chem 279:18887–18894

Hoenderop JG, van der Kemp AW, Hartog A, van de Graaf SF, van Os CH, Willems PH, Bindels RJ (1999) Molecular identification of the apical Ca^{2+} channel in 1,25-dihydroxyvitamin D3-responsive epithelia. J Biol Chem 274:8375–8378

Hofmann T, Obukhov AG, Scharfer M, Harteneck C, Gudermann T, Schultz G (1999) Direct activation of human TRPC6 and TRPC3 channels by diacyglycerol. Nature 397:259–263

Hofmann T, Schaefer M, Schultz G, Gudermann T (2002) Subunit composition of mammalian transient receptor potential channels in living cells. Proc Natl Acad Sci USA 99:7461–7466

Hoth M, Penner R (1991) Depletion of Intracellular Calcium Stores Activates a Calcium Current in Mast Cells. Nature 355, 353–356

Hotta Y, Benzer S (1970) Genetic dissection of the *Drosophila* nervous system by means of mosaics. Proc Natl Acad Sci USA 67:1156–1163

Hu Y, Schilling WP (1995) Receptor-mediated activation of recombinant *trpl* expressed in Sf9 insect cells. Biochem J 305:605–611

Hu Y, Vaca L, Zhu X, Birnbaumer L, Kunze D, Schilling WP (1994) Appearance of a novel Ca^{2+}-influx pathway in Sf9 insect cells following expression of the transient receptor potential-like (*trpl*) protein of Drosophila. Biochem Biophys Res Commun 132:346–354

Huber A, Sander P, Gobert A, Baehner M, Hermann R, Paulsen R (1996) The transient receptor potential protein (Trp), a putative store-operated Ca^{2+} channel essential for phosphoinositide-mediated photoreception, forms a signaling complex with Norp A, Ina C, InaD. EMBO J 15:7036–7045

Hunter JJ, Shao J, Smutko JS, Dussault BJ, Nagle DL, Woolf EA, Holmgren LM, Moore KJ, Shyjan AW (1998) Chromosomal localization and genomic characterization of the melastatin gene (*Mlsn 1*). Genomics 54:116–123

Inoue R, Okada T, Onoue H, Hara Y, Shimizu S, Naitoh S, Ito Y, Mori Y (2001) The transient receptor potential protein homologue TRP6 is the essential component of vascular α1-adrenoceptor-activated Ca^{2+}-permeable cation channel. Circ Res 88:325–332

Igishi T, Gutkind JS (1998) Tyrosine kinases of the *src* family participate in signaling to MAP kinase from both Gq- and Gi-coupled receptors. Biochem Biophys Res Commun 244:5–10

Jung S, Muehle A, Schaefer M, Strotmann R, Schultz G, Plant TD (2003) Lanthanides potentiate TRPC5 currents by an action at extracellular sites close to the pore mouth. J Biol Chem 278:3562–3571

Jungnickel MS, Marrero H, Birnbaumer L, Lemos JR, Florman HM (2001) Trp2 regulates Ca^{2+} entry into sperm triggered by egg ZP3. Nature Cell Biol 3:499–502

Kanzaki M, Zhang Y-Q, Mashima H, Li L, Shibata H, Kojima I (1999) Translocation of a calcium-permeable cation channel induced by insulin-like growth factor-I. Nature Cell Biol 1:165–170

Kawasaki BT, Liao Y, Birnbaumer L (2006) Molecular basis for *Src* dependence in the activation of TRPC3. Evidence for a heterogeneous makeup of ROCE and SOCE channels. Proc Natl Acad Sci USA 103:335–340

Kim JY, Saffen D (2005) Activation of M1 muscarinic acetylcholine receptors stimulates the formation of a multiprotein complex centered on TRPC6 channels. J Biol Chem 280:32035–32047

Kwan CY, Takemura H, Obie JF, Thastrup O, Putney JW Jr (1990) Effects of MeCh, Thapsigargin, and La^{3+} on plasmalemmal and intracellular Ca^{2+} transport in lacrimal acinar cells. Am J Physiol 258:C1006–1015

Kwan YK, Allchorne AJ, Vollrath MA, Christensen AP, Zhang DS, Woolf CJ, Corey DP (2006) TRPA1 contributes to cold, mechanical, and chemical nociception but is not essential for hair-cell transduction. Neuron 50: 277–289

Launay P, Fleig A, Perraud AL, Scharenberg AM, Penner R, Kinet JP (2002) TRPM4 is a Ca^{2+}-activated nonselective cation channel mediating cell membrane depolarization. Cell 109:397–407

Lee K-M, Villereal ML (1996) Tyrosine phosphorylation and activation of pp60^{c-src} and pp125^{FAK} in bradikynin stimulated fibroblasts. Am J Physiol 270:C1430-C1437

Lee K-M, Toscas K, Villereal ML (1993) Inhibition of bradikynin and thapsigargin-induced calcium entry tyrosine kinase inhibitors. J Biol Chem 268:9945–9948

Li Y, Jia YC, Cui K, Li N, Zheng ZY, Wang YZ, Yuan XB (2005) Essential role of TRPC channels in the guidance of nerve growth cones by brain-derived neurotrophic factor. Nature 434:894–898

Lievremont JP, Bird GS, Putney JW Jr (2004) Canonical transient receptor potential TRPC7 can function as both a receptor and store-operated channel in HEK-293 cells. Am J Physiol Cell Physiol 287:C1709-C1716

Liman ER, Corey DP, Dulac C (1999) TRP2, a candidate transduction channel for mammalian pheromone sensory signaling. Proc Natl Acad Sci USA 96:5791–5796

Lintschinger B, Balzer-Geldsetzer M, Baskaran T, Graier WF, Romanin C, Zhu MX, Groschner K (2000) Coassembly of Trp1 and Trp3 proteins generates diacylglycerol- and Ca^{2+}-sensitive cation channels. J Biol Chem 275:27799–277805

Liou L, Kim ML, Jones JT, Myers JW, Ferrel Jr JE, Meyer T (2005) STIM is a Ca^{2+} sensor essential for Ca^{2+}-store-depletion-triggered Ca^{2+} influx. Curr Biol 15:1235–1241

Liu X, Wang W, Singh BB, Lockwich T, Jadlowiec J, O'Connell B, Wellner R, Zhu MX, Ambudkar IS(2000) Trp1, a candidate protein for the store-operated Ca^{2+} influx mechanism in salivary gland cells. J Biol Chem 275:3403–3411

Lucas P, Ukhanov K, Leinders-Zufall T, Zufall F (2003) A diacylglycerol-gated cation channel in vomeronasal neuron dendrites is impaired in TRPC2 mutant mice: mechanism of pheromone transduction. Neuron 40:551–561

McKemy DD, Neuhausser WM, Julius D (2002) Identification of a cold receptor reveals a general role for TRP channels in thermosensation. Nature 416:52–58

Montell (2001) "Physiology, Phylogeny and Functions of the TRP Superfamily of Cation Channels." Science's SKTE http://skte.sciencemag.org/cgi/content/full/OC_sigtrans;2001/90/re1

Montell C, Rubin GM (1989) Molecular characterization of the Drosophila *trp* locus: a putative integral membrane protein required for phototransduction. Neuron 2:1313–1323

Montell C, Birnbaumer L, Flockerzi V, Bindels RJ, Caterina MJ, Clapham DE, Heller S, Julius D, Scharenberg AM, Schultz G, Zhu MX (2002) A unified nomenclature for the superfamily of TRP cation channels. Mol Cell 9:229–231

Nadler MJ, Hermosura MC, Inabe K, Perraud AL, Zhu Q, stokes AJ, Kurosaki T, Kinet JP, Penner R, Scharenberg AM, Flieg A (2001) LTRPC7 is a MgATP-regulated divalent cation channel required for cell viability. Nature 411:590–595

Okada T, Inoue R, Yamazaki K, Maeda A, Kurosaki T, Yamakuni T, Tanaka I, Shimizu S, Ikenaka K, Imoto K, Mori Y (1999) Molecular and functional characterization of a novel mouse TRP homologue TRP7 that forms a background and receptor-activated Ca^{2+} permeable cation channel. J Biol Chem 274:27359–27370

Pak WL, Grossfield J, Arnold K (1970) Mutant of the visual pathway of *Drosophila melanogaster*. Nature 227:518–520

Panda S, Nayak SK, Campo B, Walker JR, Hogenesch JB, Jegla T (2005) Illumination of the melanopsin signaling pathway. Science 307:600–604

Patterson RL, van Rossum DB, Gill DL (1999) Store operated Ca^{2+} entry: evidence for a secretion-like coupling model. Cell 98:487–499

Peng JB, Chen XZ, Berger UV, Vassilev PM, Tsukaguchi H, Brown EM, Hediger MA (1999) Molecular cloning and characterization of a channel-like transporter mediating intestinal calcium absorption. J Biol Chem 274:22739–22746

Perez CA, Huang L, Rong M, Kozak, JA, Preuss AK, Zhang H, Max M, Margolskee RF (2002) A transient receptor potential channel expressed in taste receptor cells. Nat Neurosci 5:1169–1176

Perraud AL, Fleig A, Dunn CA, Bagley LA, Launay P, Schimitz C, Stokes AJ, Zhu Q, Bessman MJ, Penner R, Kinet JP, Scharenberg AM (2001) ADP-ribose gating of the calcium-permeable LTRPC2 channel revealed by Nudix motif homology. Nature 411:595–599

Philipp S, Cavalie A, Freichel M, Wissenbach U, Zimmer S, Trost C, Marquart A, Murakami M, Flockerzi V (1996) A mammalian capacitative calcium entry channel homologous to Drosophila TRP, TRPL. EMBO J 25:6166–6171

Philipp S, Hambrecht J, Braslavski L, Schroth G, Freichel M, Murakami M, Cavalie A, Flockerzi V (1998) A novel capacitative calcium entry channel expressed in excitable cells. EMBO J 77:4274–4282

Phillips AM, Bull A, Kelly LE (1992) Identification of a Drosophila gene encoding a calmodulin-binding protein with homology to the *trp* phototransduction gene. Neuron 8:631–642

Putney JW Jr (1986) A model for receptor-regulated calcium entry. Cell Calcium 7:1–12

Putney JW Jr (1990) Capacitative calcium entry revisited. Cell Calcium 11:611–624

Qiu X, Kumbalasiri T, Carlson SM, Wong KY, Krishna V, Provencio I, Berson DM (2005) Induction of photosensitivity by heterologous expression of melanopsin. Nature 433:745–749

Riccio A, Medhurst AD, Mattei C, Kelsell RE, Calver AR, Randall AD, Benham CD, Pangalos MN (2002a) mRNA distribution analysis of human TRPC family in the CNS and peripheral tissues. Mol Brain Res 109:v95–104

Riccio A, Mattei C, Kelsell RE, Medhurst AD, Calver AR, Randall AD, Davis JD, Benham CD, Pangalos MN (2002b) Cloning and Functional Expression of Human Short TRP7, a Candidate Protein for Store-operated Ca^{2+} Influx. J Biol Chem 277:12302–12309

Rios E, Pizarro G (1991) Voltage sensor of excitation-contraction coupling in skeletal muscle. Physiol Rev 71:849–908

Roos J, DiGregorio PJ, Yeromin AV, Ohlsen K, Lioudyno M, Zhang S, Safrina O, Kozak JA, Wagner SL, Cahalan MD, Vilicelebi G, Stauderman KA (2005) STIM1, an essential and conserved component of store-operated Ca^{2+} channel function. J Cell Biol 169:435–445

Rosado JA, Sage SO (2001) Activation of store-mediated calcium entry by secretion-like coupling between the inositol 1,4,5-trisphosphate receptor type II and human transient receptor potential (hTrp1) channels in human platelets. Biochem J 356:191–198

Rosker C, Graziani A, Lukas M, Eder P, Zhu MX, Romanin C, Groschner K (2004) Ca^{2+} signaling by TRPC3 involves Na^+ entry and local coupling to the Na^+/Ca^{2+} exchanger. J Biol Chem 279:13696–13704

Schaefer M, Plant TD, Stresow N, Albrecht N, Schultz G (2002) Functional differences between TRPC4 splice variants. J Biol Chem 277:3752–3759

Schindl R, Kahr H, Graz I, Groschner K, Romanin C (2002) Store depletion-activated CaT1 currents in rat basophilic leukemia mast cells are inhibited by 2-aminoethoxydiphenyl borate. Evidence for a regulatory component that controls activation of both CaT1 and CRAC (Ca^{2+} release-activated Ca^{2+} channel) channels. J Biol Chem 277:26950–26958

Selinger Z, Minke B (1988) Inositol lipid cascade of vision studied in mutant flies. Cold Spring Harbor Symp Quant Biol 53:333–341

Shieh B-H, Zhu M-Y (1995) Regulation of the TRP Ca^{2+} channel by INAD in Drosophila photoreceptors. Neuron 16:991–998

Sinkins WG, Vaca L, Hu Y, Kunze DL, Schilling WP (1996) The COOH-terminal domain of Drosophila TRP channels confers Thapsigargin sensitivity. J Biol Chem 271:2955–2960

Soboloff J, Spassova M, Xu W, He LP, Cuesta1 N, Gill DL (2005) Role of endogenous TRPC6 channels in Ca^{2+} signal generation in A7r5 smooth muscle cells. J Biol Chem 280:39786–39794

Story GM, Peier AM, Reeve AJ, Eid SR, Mosbacher J, Hricik TR, Earley TJ, Hergarden HC, Andersson DA, Hwang SW, McIntyre, Jegla T, Bevan S, Patapoutian A (2003) ANKTM1, a TRP-like channel expressed in nociceptive neurons, is activated by cold temperatures. Cell 112:819–829

Stowers L, Holy TE, Meister M, Dulac C, Koentges G (2002) Loss of sex discrimination and male-male aggression in mice deficient for TRPC2. Science 295:1493–1500

Strotmann R, Harteneck C, Nunnenmacher K, Schultz G, Plant TD (2000) OTRPC4, a non-selective cation channel that confers sensitivity to extracellular osmolarity. Nat Cell Biol 2:695–702

Struebing C, Krapivinsky G, Krapivinsky L, Clapham DE (2003) Formation of novel TRPC channels by complex subunit interactions in embryonic brain. J Biol Chem 278:39014–39019

Sun M, Goldin E, Stahl S, Falardeau JL, Kennedy JC, Acierno JS, Bove C, Kaneski CR, Nagel J, Bromley MC, Colman M, Schiffmann R Slaugenhaupt SA (2000) Mucolipidosis type 4 is caused by mutations in a gene encoding a novel transient receptor potential channel. Human Mol Genet 9:2471–2478

Tang Y, Tang J, Chen Z, Trost C, Flockerzi V, Li M, Rameshi V, Zhu MX (2000) Association of mammalian Trp4 and phospholipase C isozymes with a PDZ domain-containing protein, NHERF. J Biol Chem 275:37559–37564

Thastrup O, Cullen PJ, Drobak BK, Hanley MR, Dawson AP (1990) Thapsigargin, a tumor promoter, discharges intracellular Ca^{2+} stores by specific inhibition of the endoplasmic reticulum Ca^{2+} ATPase. Proc Natl Acad Sci USA 87:2466–2470

Trebak M, Hempel N, Wedel BJ, Smyth JT, Bird GS, Putney JW Jr. (2005) Negative regulation of TRPC3 channels by protein kinase C-mediated phosphorylation of serine 712. Mol Pharmacol 67:558–563

Vannier B, Peyton M, Boulay G, Brown D, Qin N, Jiang M, Zhu X, Birnbaumer L (1999) Mouse trp2, the homologue of the human trpc2 pseudogene, encodes mTrp2, a store depletion-activated capactitative Ca^{2+} entry channel. Proc Natl Acad Sci USA 96:2060–2064

Vazquez G, Wedel BJ, Trebak M, Bird GS, Putney JW Jr. (2003) Expression level of the canonical transient receptor potential 3 (TRPC3) channel determines its mechanism of activation. J Biol Chem 278:21649–21654

Vazquez G, Wedel BJ, Kawasaki BT, St. John Bird G, Putney J (2004) Obligatory role for *src* kinase in the signaling mechanism for TRPC3 cation channels. J Biol Chem 279:40521–40528

Veldhuisen B, Spruit L, Dauwerse HG, Breuning MH, Peters DJM (1999) Genes homologous to the autosomal dominant polycystic kidney disease genes (PKD1 and PKD2). Eur J Human Genet 7:860–872

Venkatachalam K, Zheng F, Gill DL (2003) Regulation of canonical transient receptor potential (TRPC) channel function by diacylglycerol and protein kinase C. J Biol Chem 278:29031–29040

Wang GX, Poo MM (2005) Requirement of TRPC channels in netrin-1-induced chemotropic turning of nerve growth cones. Nature 434:898–904

Wes PD, Chevesich J, Jeromin A, Rosenberg C, Stetten G, Montell C (1995) TRPC1, a human homolog of a *Drosophila* store-operated channel. Proc Natl Acad Sci USA 92:9652–9656

Xu X-ZS, Choudhury A, Li X, Montell C (1998) Coordination of an array of signaling proteins through homo- and heteromeric interactions between PDZ domains and target proteins. J Cell Biol 142:545–555

Yildirim E, Kawasaki BT, Birnbaumer L (2005) Molecular cloning of TRPC3a, an N-terminally extended, store-operated variant of the human C3 transient receptor potential channel. Proc Natl Acad Sci USA 102:3307–3311

Yue L, Peng JB, Hediger MA, Clapham DE (2001) CaT1 manifests the pore properties of the calcium release activated calcium channel. Nature 410:705–709

Zagranichnaya TK, Wu X, Villereal ML (2005) Endogenous TRPC1, TRPC3, and TRPC7 proteins combine to form native store-operated channels in HEK-293 cells. J Biol Chem 280:29559–29569

Zhang SL, Yu Y, Roos J, Kozak JA, Deerinck TJ, Ellisman MH, Stauderman KA, Cahalan MD (2005) STIM1 is a Ca^{2+} sensor that activates CRAC channels and migrates from the Ca^{2+} store to the plasma membrane. Nature 437:902–905

Zhu X, Chu PB, Peyton M, Birnbaumer L (1995) Molecular cloning of a widely expressed human homologue for the *Drosophila trp* gene. FEBS Lett 373:193–198

Zhu X, Jiang M, Peyton MJ, Boulay G, Hurst R, Stefani E, Birnbaumer L (1996) *trp*, a novel mammalian gene family essential for agonist-activated capacitative Ca^{2+} influx. Cell 85:661–671

Zufall F, Ukhanov K, Lucas P, Leinders-Zufall T (2005) TRPC2: from gene to behavior. Pfluegers Arch Eur J Physiol 451:61–71

Zweifach A, Lewis RS (1993) Mitogen-regulated Ca^{2+} current of T lymphocytes is activated by depletion of intracellular Ca^{2+} stores. Proc Natl Acad Sci USA 90:6295–6299

Functional Rescue of Misfolded Receptor Mutants

Shaun P. Brothers[1] and *P. Michael Conn*[1,2]

Summary

Mutant gonadotropin-releasing hormone (GnRH) receptors isolated from patients with GnRH-resistant hypogonadotropic hypogonadism are frequently proteins that are misrouted in the cell. Such mutant receptors are retained in the endoplasmic reticulum and can be rescued by pharmacological chaperones. This understanding contrasts with the view that these mutant receptors lose the ability to bind ligand or effect signal transduction. Pharmacological chaperones, or "pharmacoperones," bind specifically to GnRH receptors and allow them to escape retention by the cellular quality control systems and route to the plasma membrane, where they function normally. This observation suggests that pharmacoperones have the potential to be used to treat a number of human diseases characterized by misrouted proteins, among these, hypogonadotropic hypogonadism, cystic fibrosis and nephrogenic diabetes insipidus.

Introduction

Gonadotropin-releasing hormone (GnRH), a hormone secreted by hypothalamic neurons, travels through a portal blood system to anterior pituitary gonadotrope cells where it binds to the GnRH receptor (GnRHR) at the plasma membrane. The GnRHR is a G protein coupled receptor (GPCR), and agonist activation of the GnRHR leads to initiation of intracellular signaling cascades, a major physiological consequence of which is the synthesis and release of the gonadotropin hormones, luteinizing hormone (LH) and follicle-stimulating hormone (FSH). LH and FSH enter the blood stream, then stimulate steroidogenesis and regulate reproductive function. GnRH signaling is, of necessity, highly regulated, and the amplitude as well as frequency of GnRH release convey information (Wildt et al. 1981; Filicori et al. 1986). Besides regulation of GnRH release by neural inputs and steroid and glycoprotein hormones, GnRHR expression also appears to be highly regulated at the transcriptional and post-translational levels (Schroder and Kaufman 2005). This intricate control is advantageous when controlling the complicated human reproductive cycle but becomes disadvantageous when mutations are introduced. Here we discuss a cellular mechanism causing GnRH-resistant hypogonadotropic hypogonadism, a phenotype that can potentially be reversed using pharmacologic intervention; we also discuss approaches of receptor rescue with pharmacological chaperones or "pharmacoperones".

[1] Divisions of Neuroscience and [2]Reproductive Biology, Oregon National Primate Research Center and Departments of [1]Physiology and Pharmacology and [2]Cell and Developmental Biology, Oregon Health and Science University

Conn et al.
Insights into Receptor Function
and New Drug Development Targets
© Springer-Verlag Berlin Heidelberg 2006

The GnRHR in Reproduction

Gonadotropin-releasing hormone (GnRH) is a decapeptide (pGlu-His-Trp-Ser-Tyr-Gly-Leu-Arg-Pro-Gly-NH$_2$) produced by hypothalamic neurons. The discovery and characterization of GnRH resulted in a Nobel Prize in Medicine awarded to Andrew Schally and Roger Guillemin in 1977. The interaction of GnRH with its cognate receptor, and subsequent release of the gonadotropins, is the interface between the brain and gonadal function (Fig. 1).

The GnRHR is a seven transmembrane, G protein coupled receptor (GPCR) belonging to the rhodopsin-like family of these receptors (Fig. 2; Sealfon 2005). Agonist activation of the GnRHR results in activation of the GTP-binding protein, Gq. The active form of Gq initiates an intracellular signaling cascade of molecules by switching on the enzyme phospholipase C (PLC). The integral membrane molecule phosphatidyl inositol 3,4,5 triphosphate is cleaved by PLC into diacylglycerol (DAG) and inositol 3,4,5 triphosphate (IP$_3$). The presence of both of these second messengers activate protein

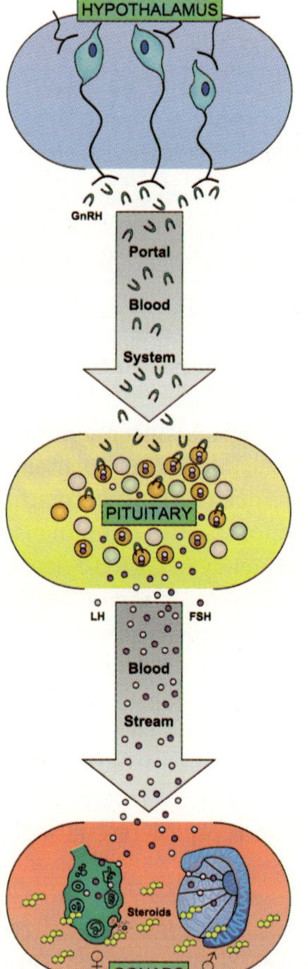

Fig. 1. The hypothalamic-pituitary-gonadal axis. Hypothalamic GnRH is released and travels through the portal blood system to the anterior pituitary, where gonadotrope cells expressing the GnRH receptor are stimulated and synthesize and release LH and FSH. The gonadotropins are released in the blood stream and act on reproductive tissues to initiate steroidogenesis and gametogenesis

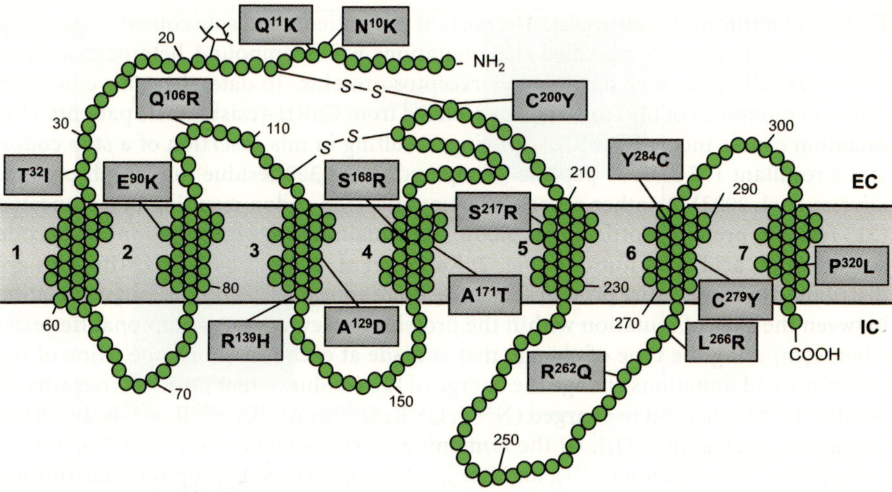

Fig. 2. The GnRH receptor is a seven transmembrane segment, G protein coupled receptor. The human GnRH receptor protein is made up of 328 amino acids. Shown are the 16 single amino acid residue substitutions that have been isolated from patients with hypogonadotropic hypogonadism. Not shown are the two mutations isolated from HH patients that result in truncated forms of the GnRH receptor, those being 177 and 313 amino acids in length

kinase C (PKC) and other downstream kinases, signaling molecules and transcription factors, resulting in the synthesis and release of LH and FSH into the blood stream, where they travel to the gonads to initiate steroidogenesis and gametogenesis (for an extensive review of GnRHR intracellular signaling pathways, see Cheng and Leung 2005).

Hypogonadotropic Hypogonadism

Hypogonadotropic hypogonadism (HH) is characterized by delayed or absent sexual development due to partial or complete loss of circulating gonadotropin and/or sex steroid levels (Karges et al. 2003). Adult HH patients also display a range of other phenotypic traits, including low testicular volume, microphallus (in men), primary amenorrhoea (in women) and the absence, in both sexes, of axillary and/or pubic hair (Karges et al. 2003). Typically, steroid hormone replacement therapy is used to treat HH patients (Karges et al. 2003). There are several etiologies of hypogonadotropic hypogonadism, all of which can be traced to loss of integrity in the hypothalamic-pituitary-gonadal axis (Iovane et al. 2004). Among the most studied HH etiologies are the loss of GnRH neuron migration from the olfactory placode (also associated with anosmia), lesions of the portal blood vasculature by trauma or tumor (also associated with adrenal insufficiency), loss of gonadotropin signaling capability and loss of GnRH signaling capability (for a review of these etiologies, see Karges et al. 2003).

This last etiology was identified in patients who were diagnosed with HH but were unresponsive to treatment with exogenous GnRH analogs (de Roux et al. 1997).

changes to the receptor, such as the deletion of a single amino acid at position 191 in the receptor protein sequence, also had a similar effect (Janovick et al. 2003). Furthermore, when these same changes were made to an HH mutant receptor, rescue of the protein was affected as well (Maya-Nunez et al. 2002; Brothers et al. 2003). Protein modifications, such as epitope tagging or residue deletion in order to affect rescue, are of limited value outside of academic considerations because, after modification, the original protein sequence has been altered.

Rescue of unaltered proteins, normally retained inside the cell, has been affected by using small molecules known as pharmacological chaperones or "pharmacoperones" (Ulloa-Aguirre et al. 2004; Castro-Fernandez et al. 2005). Pharmacoperones for the GnRH receptor are high affinity receptor antagonists that are competitive with the natural ligand and hydrophobic enough to be membrane permeable. IN3, the pharmacoperone most widely used to rescue the GnRH receptor (Fig. 4), is a peptidomimetic molecule that was synthesized by Wallace T. Ashton and Mark Goulet at Merck and Co. (Ashton et al. 2001).

When IN3 was introduced to cells that expressed the wild type GnRHR for a short time, then removed, and such cells were stimulated with GnRHR agonist, the signaling

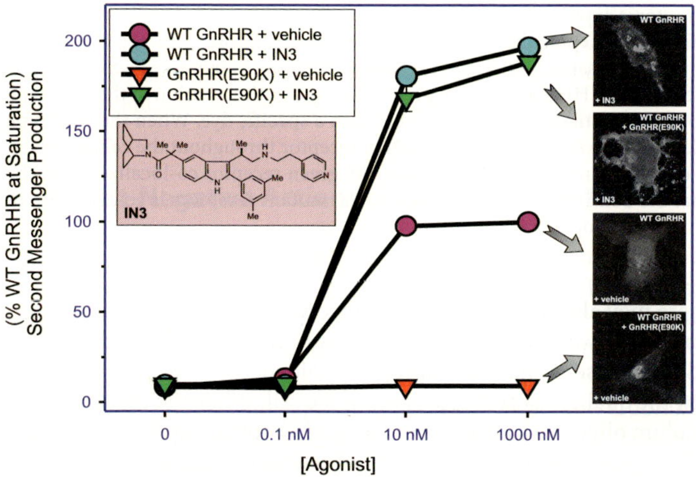

Fig. 4. The pharmacological chaperone, IN3, and its action on the GnRH receptor. When cells expressing the wild type GnRHR are treated with IN3, then the pharmacoperone is removed and the cells are stimulated with receptor agonist, there is increased second messenger production compared to cells treated with only vehicle. Similarly, in cells expressing the mutant receptors isolated from HH patients, such as the $E^{90}K$ receptor, after treatment with IN3, there is a vast increase in the second messenger production after agonist treatment compared to cells treated with only vehicle. Inset micrographs: Cells expressing the GFP-chimer of the wild type GnRHR alone, or with the dominant-negative $E^{90}K$ mutant, were also treated with pharmacological chaperone and then imaged with a confocal microscope. The wild type receptor is localized to the plasma membrane after pharmacoperone treatment, and the cells that express the dominant-negative mutant also show significant redistribution of the wild type receptor to the plasma membrane. The chemical structure of IN3 ((2S)-2-[5-[2-(2-azabicyclo[2.2.2]oct-2-yl)-1,1-dimethyl-2-oxoethyl]-2-(3,5-dimethylphenyl)-1H-indol-3-yl]-N-(2-pyridin-4-ylethyl) propan-1-amine) is inset as shown

in those cells increased dramatically (Fig. 4; Janovick et al. 2002). Additionally, when IN3 was administered to cells expressing the mutant receptors isolated from HH patients, a majority of those receptors were also rescued and the dominant-negative effect of those rescued receptors was ablated (Leanos-Miranda et al. 2003). Only 2 of the 16 single site mutants continued to have a dominant-negative effect on the wild type receptor (Janovick et al. 2002). When confocal microscopy was used to visualize rescued and unrescuable receptors, it confirmed that the rescued receptors did, indeed, escape the endoplasmic reticulum and route to the plasma membrane (Fig. 4), whereas the unrescuable ones remained retained in the endoplasmic reticulum (Brothers et al. 2004). Significantly, it appears as though most of the mutant proteins were fully capable of binding to ligand and transducing a signal, but the receptors were simply misrouted within the cell and so were not accessible to ligand. The receptors that were rescued to the plasma membrane appeared to function identically to the wild type receptor, with the same rank order of potency for ligands, the same affinity for agonist and the same turnover rates (Janovick et al. 2002). The only difference that could be observed was that there was little or no plasma membrane expression in the absence of pharmacoperone.

General Applicability

While the GnRH receptor is among the most studied proteins that can be rescued by pharmacoperones, it is by no means unique. A number of diseases are caused by mutation and resultant misrouting (Castro-Fernandez et al. 2005). They include cystic fibrosis [cystic fibrosis transmembrane conductance regulator (CFTR) chloride channel], retinitis pigmentosa (rhodopsin) and nephrogenic diabetes insipidus (aquaporin 2 channel and V2 vasopressin receptor; Table 1). Pharmacoperone treatment of all of these proteins has been considered. For the V2-vasopressin receptor, several well-characterized pharmacoperones have been reported to affect rescue of mutant proteins found in nephrogenic diabetes insipidus patients (Morello et al. 2000). Mutant CFTR chloride channels, including the CFTR(del^{508}) mutant found in a majority of cystic fibrosis patients, have been rescued using non-specific chaperones, such as dimethylsulfoxide and glycerol, that change the solvent properties in the endoplasmic reticulum to make the energetics of folding more favorable (Lim and Zeitlin 2001).

In addition to protein mutation causing misrouting, unwanted aggregation of mutated proteins can also result (Castro-Fernandez et al. 2005). In Alzheimer's and Parkinson's diseases, the mutant proteins form aggregates in the cell and slowly lead to degenerative function of those cells (Castro-Fernandez et al. 2005; Muchowski and Wacker 2005). Prophylactic pharmacoperone treatment that would prevent such proteins from misfolding into a conformation that is suitable for inclusion into a protein crystal has been considered for these and other protein aggregation diseases (Table 1; Castro-Fernandez et al. 2005). The treatment would likely have to be prophylactic, since the energetics of solvating a protein crystal or dissolving an aggregate in vivo are quite unfavorable. Prophylaxis could be administered to high-risk patients early enough to prevent or slow the progression of the disease.

The proper folding and successful routing of the many thousands of proteins synthesized by the cell is dependent on the stringent quality control mechanism in the endoplasmic reticulum and Golgi (Sitia and Braakman 2003). This quality control is

Table 1. Several diseases caused by misfolded or aggregated proteins and corresponding rescue agents

Protein Anomaly	Disease	Causative Protein(s)	Rescue Agents	References
Misrouting:	Cystic fibrosis	CFTR	DMSO, glycerol, xanthines	Lim and Zeitlin 2001
	Systemic amyloidosis	Amyloid fibriis	β-sheet breaking peptides	Soto et al. 2000 Permanne et al. 2002
	Nephrogenic diabetes insipidus	Aquaporin-2, V2-vasopressin receptor	Glycerol, triethylamines peptidomimetic receotor antagonists	Tamarappoo and Verkman 1998 Morello et al. 2000
	Cancer	p53	Alkaloids, glycerol, triethylamines	Peng et al. 2003 Fiedler et al. 2003
	Hypogonadotropic hypogonadism	GnRH receptor	Peptidomimetic receptor antagonists	Janovik et al. 2002
	Retinitis pigmentosa	Rhodopsin, carotenoid receptor	11-cys-7-ring-retinals	Noorwerz et al. 2004
	Emphysema	α1-antitrypsin	4-phenylbuteric acid	Burrows et al. 2000
	α1-antitrypsin deficiency liver disease	α1-antitrysin	4-phenylbuteric acid	Burrows et al. 2000
Aggregation:	Alzheimer's disease	Amyloid, tau	β-sheet breaking peptides	Soto et al. 2000 Permanne et al. 2002
	Cataracts	Crystalins	Phase separation inhibitors	Benedek et al. 1999
	Creutzfeldt-Jacob disease	Amyloid	β-sheet breaking peptides	Soto et al. 2000 Permanne et al. 2002
	Huntington's disease	Huntington	Thioflavins, chrysamine G	Heiser et al. 2000
	Parkinson's disease	α-synuclein, parkin	–	Forloni et al. 2002
	Spongioform enecephalopathy	Prions	β-sheet breaking peptides	Soto et al. 2000 Permanne et al. 2002
	Sickle-cell anemia	Hemoglobin	[K+] increasing agents	Brugnara 2003

necessarily non-specific so that the folded state of each protein can be determined. Generally, recognition of exposed hydrophobic surfaces on proteins signals that a protein is misfolded and will be targeted for re-folding or degradation (Sitia and Braakman 2003). Molecular chaperones such as calnexin/calreticulin and many of the heat shock protein family act as the endoplasmic reticulum folding sensors and direct misfolded proteins to re-folding pathways, to their functional site in the cell or to degradation pathways (Schroder and Kaufman 2005). Likely, pharmacoperones act to help the protein fold into a state where quality control degradation of mutant proteins no longer occurs, as is the case with the CFTR mutant proteins (Kerem 2005). Understanding the underlying basis of protein folding and routing, as well as nascent protein interaction with the intrinsic quality control system, will likely provide a range of novel therapeutic approaches to treat a number of human diseases.

Concluding Remarks

Given the increasing numbers of misfolded proteins that are misrouted and cause disease, there will likely be many prospects for pharmacologic reversal of such misfolding and resultant misrouting. However, a non-specific pharmacological agent to decrease retention and degradation of misfolded proteins is, perhaps, not the wisest choice because proteins that are properly formed but retained as a part of their normal regulation would also be rescued. There are a few desirable characteristics of pharmacoperones that would increase specificity and ensure the best possible result of treatment. Ideally, the pharmacoperone would be highly selective for the protein, it would be membrane permeable, it would be able to get to the site of action (i.e., pass the blood-brain barrier), and it would ideally be easily administered to the patient.

With the ever-increasing number of diseases found to have a defect in protein folding and/or routing, pharmacoperone restoration of the protein native state is becoming an ever-more attractive idea. Industry and academic scientists alike are pursuing avenues of research that will lead to compounds that serve as folding templates and restore functionality to proteins such as the GnRH receptor in hypogonadotropic hypogonadism and the CFTR in cystic fibrosis (for a recent review of these and other protein folding diseases, see Castro-Fernandez et al. 2005).

Acknowledgements. Supported by HD-19899, RR-00163 and HD-18185

References

Ashton WT, Sisco RM, Yang YT, Lo JL, Yudkovitz JB, Gibbons PH, Mount GR, Ren RN, Butler BS, Cheng K, Goulet MT (2001) Potent nonpeptide GnRH receptor antagonists derived from substituted indole-5-carboxamides and -acetamides bearing a pyridine side-chain terminus. Bioorg Med Chem Lett 11:1727–1731

Benedek GB, Pande J, Thurston GM, Clark JI (1999) Theoretical and experimental basis for the inhibition of cataract. Prog Retin Eye Res 18:391–402

Brothers SP, Janovick JA, Conn PM (2003) Unexpected effects of epitope and chimeric tags on gonadotropin-releasing hormone receptors: implications for understanding the molecular etiology of hypogonadotropic hypogonadism. J Clin Endocrinol Metab 88:6107–6112

Brothers SP, Cornea A, Janovick JA, Conn PM (2004) Human loss-of-function gonadotropin-releasing hormone receptor mutants retain wild-type receptors in the endoplasmic reticulum: molecular basis of the dominant-negative effect. Mol Endocrinol 18:1787–1797

Brugnara C (2003) Sickle cell disease: from membrane pathophysiology to novel therapies for prevention of erythrocyte dehydration. J Pediatr Hematol Oncol 25:927–933

Burrows JA, Willis LK, Perlmutter DH (2000) Chemical chaperones mediate increased secretion of mutant alpha 1-antitrypsin (alpha 1-AT) Z: A potential pharmacological strategy for prevention of liver injury and emphysema in alpha 1-AT deficiency. Proc Natl Acad Sci USA 97:1796–1801

Castro-Fernandez C, Maya-Nunez G, Conn PM (2005) Beyond the signal sequence: protein routing in health and disease. Endocr Rev 26:479–503

Cheng CK, Leung PC (2005) Molecular biology of gonadotropin-releasing hormone (GnRH)-I, GnR H-II, and their receptors in humans. Endocr Rev 26:283–306

Conn PM, Rogers DC, Stewart JM, Niedel J, Sheffield T (1982) Conversion of a gonadotropin-releasing hormone antagonist to an agonist. Nature 296:653–655

Cornea A, Janovick JA, Maya-Nunez G, Conn PM (2001) Gonadotropin-releasing hormone receptor microaggregation. Rate monitored by fluorescence resonance energy transfer. J Biol Chem 276:2153–2158

de Roux N, Young J, Misrahi M, Genet R, Chanson P, Schaison G, Milgrom E (1997) A family with hypogonadotropic hypogonadism and mutations in the gonadotropin-releasing hormone receptor. N Engl J Med 337:1597–1602

Filicori M, Santoro N, Merriam GR, Crowley WF Jr (1986) Characterization of the physiological pattern of episodic gonadotropin secretion throughout the human menstrual cycle. J Clin Endocrinol Metab 62:1136–1144

Forloni G, Terreni L, Bertani I, Fogliarino S, Invernizzi R, Assini A, Ribizzi G, Negro A, Calabrese E, Volonte MA, Mariani C, Franceschi M, Tabaton M, Bertoli A (2002) Protein misfolding in Alzheimer's and Parkinson's disease: genetics and molecular mechanisms. Neurobiol Aging 23:957–976

Friedler A, DeDecker BS, Freund SM, Blair C, Rudiger S, Fersht AR (2004) Structural distortion of p53 by the mutation R249S and its rescue by a designed peptide: implications for "mutant conformation". J Mol Biol 336:187–196

Heiser V, Scherzinger E, Boeddrich A, Nordhoff E, Lurz R, Schugardt N, Lehrach H, Wanker EE (2000) Inhibition of huntingtin fibrillogenesis by specific antibodies and small molecules: implications for Huntington's disease therapy. Proc Natl Acad Sci USA 97:6739–6744

Iovane A, Aumas C, de Roux N (2004) New insights in the genetics of isolated hypogonadotropic hypogonadism. Eur J Endocrinol 151:U83–88

Janovick JA, Maya-Nunez G, Conn PM (2002) Rescue of hypogonadotropic hypogonadism-causing and manufactured GnRH receptor mutants by a specific protein-folding template: misrouted proteins as a novel disease etiology and therapeutic target. J Clin Endocrinol Metab 87:3255–3262

Janovick JA, Ulloa-Aguirre A, Conn PM (2003) Evolved regulation of gonadotropin-releasing hormone receptor cell surface expression. Endocrine 22:317–327

Karges B, Karges W, de Roux N (2003) Clinical and molecular genetics of the human GnRH receptor. Human Reprod Update 9:523–530

Kerem E (2005) Pharmacological induction of CFTR function in patients with cystic fibrosis: mutation-specific therapy. Pediatr Pulmonol 40:183–196

Kottler ML, Chauvin S, Lahlou N, Harris CE, Johnston CJ, Lagarde JP, Bouchard P, Farid NR, Counis R (2000) A new compound heterozygous mutation of the gonadotropin-releasing hormone receptor (L314X, Q106R) in a woman with complete hypogonadotropic hypogonadism: chronic estrogen administration amplifies the gonadotropin defect. J Clin Endocrinol Metab 85:3002–3008

Leanos-Miranda A, Janovick JA, Conn PM (2002) Receptor-misrouting: an unexpectedly prevalent and rescuable etiology in gonadotropin-releasing hormone receptor-mediated hypogonadotropic hypogonadism. J Clin Endocrinol Metab 87:4825–4828

Leanos-Miranda A, Ulloa-Aguirre A, Ji TH, Janovick JA, Conn PM (2003) Dominant-negative action of disease-causing gonadotropin-releasing hormone receptor (GnRHR) mutants: a trait that potentially coevolved with decreased plasma membrane expression of GnRHR in humans. J Clin Endocrinol Metab 88:3360–3367

Lim M, Zeitlin PL (2001) Therapeutic strategies to correct malfunction of CFTR. Paediatr Respir Rev 2:159–164

Maya-Nunez G, Janovick JA, Ulloa-Aguirre A, Soderlund D, Conn PM, Mendez JP (2002) Molecular basis of hypogonadotropic hypogonadism: restoration of mutant (E(90)K) GnRH receptor function by a deletion at a distant site. J Clin Endocrinol Metab 87:2144–2149

Morello JP, Salahpour A, Laperriere A, Bernier V, Arthus MF, Lonergan M, Petaja-Repo U, Angers S, Morin D, Bichet DG, Bouvier M (2000) Pharmacological chaperones rescue cell-surface expression and function of misfolded V2 vasopressin receptor mutants. J Clin Invest 105:887–895

Muchowski PJ, Wacker JL (2005) Modulation of neurodegeneration by molecular chaperones. Nature Rev Neurosci 6:11–22

Noorwez SM, Malhotra R, McDowell JH, Smith KA, Krebs MP, Kaushal S (2004) Retinoids assist the cellular folding of the autosomal dominant retinitis pigmentosa opsin mutant P23H. J Biol Chem 279:16278–16284

Peng Y, Li C, Chen L, Sebti S, Chen J (2003) Rescue of mutant p53 transcription function by ellipticine. Oncogene 22:4478–4487

Permanne B, Adessi C, Saborio GP, Fraga S, Frossard MJ, Van Dorpe J, Dewachter I, Banks WA, Van Leuven F, Soto C (2002) Reduction of amyloid load and cerebral damage in a transgenic mouse model of Alzheimer's disease by treatment with a beta-sheet breaker peptide. FASEB J 16:860–2

Schroder M, Kaufman RJ (2005) The mammalian unfolded protein response. Annu Rev Biochem 74:739–789

Sealfon SC (2005) G protein-coupled receptors. Sci STKE 279:tr11

Silveira LF, Stewart PM, Thomas M, Clark DA, Bouloux PM, MacColl GS (2002) Novel homozygous splice acceptor site GnRH receptor (GnRHR) mutation: human GnRHR "knockout". J Clin Endocrinol Metab 87:2973–2977

Sitia R, Braakman I (2003) Quality control in the endoplasmic reticulum protein factory. Nature 426:891–894

Soto C, Kascsak RJ, Saborio GP, Aucouturier P, Wisniewski T, Prelli F, Kascsak R, Mendez E, Harris DA, Ironside J, Tagliavini F, Carp RI, Frangione B (2000) Reversion of prion protein conformational changes by synthetic beta-sheet breaker peptides. Lancet 355:192–197

Tamarappoo BK, Verkman AS (1998) Defective aquaporin-2 trafficking in nephrogenic diabetes insipidus and correction by chemical chaperones. J Clin Invest 101:2257–2267

Ulloa-Aguirre A, Janovick JA, Brothers SP, Conn PM (2004) Pharmacologic rescue of conformationally-defective proteins: implications for the treatment of human disease. Traffic 5:821–837

Wildt L, Hausler A, Marshall G, Hutchison JS, Plant TM, Belchetz PE, Knobil E (1981) Frequency and amplitude of gonadotropin-releasing hormone stimulation and gonadotropin secretion in the rhesus monkey. Endocrinology 109:376–385

Obesity-related mutations of leptin and melanocortin receptors

Cécile Lubrano[1], *Béatrice Dubern*[1], and *Karine Clément*[1]

Summary

In recent years, the molecular approach to human obesity has advanced the understanding of some causes and mechanisms of severe forms of obesity. Single rare mutations largely contribute to the development of some obesity cases. Research was conducted after a meticulous clinical evaluation of individuals with specific biochemical or hormonal anomalies. These obesity cases are very severe and generally start in childhood. This chapter will focus on genetic mutations causing primary defects in the leptin and melanocortin pathways. Although obesity due to mutations of leptin, leptin receptor, proopiomelanocortin, and proconvertase 1 are exceptional, obesity linked to MC4R mutations could represent 2 to 4% of human cases. The phenotypic and endocrine features of these mutations causing a dysfunction in leptin and melanocortin signaling will be reviewed. The contribution of genetic variations of genes encoding the key actors of the leptin and melanocortin pathways in common forms of obesity will also be discussed.

Introduction

Obesity is associated with many genetic syndromes. To identify new pathways involved in the control of body weight, several teams are attempting to identify the genes and mutations responsible for the development of severe obesity cases. The OMIM database (http://www.ncbi.nlm.nih.gov/entrez/Query.fcgi?db=OMIM) lists them and provides access to the clinical description of these rare diseases. A significant success was derived from the study of candidate genes implicated in rodent monogenic obesity. Research was conducted in obese children or young adults after a meticulous clinical evaluation comprising the description of specific biochemical or hormonal anomalies (Farooqi and O'Rahilly 2004). In these cases, the genetic anomalies affected key factors of body weight regulation linked to leptin action – the pivotal hormone controlling weight regulation and several endocrine pathways – and the melanocortin pathway, the target of the leptin in the hypothalamus (Fig. 1).

These studies have contributed to validate a crucial role of the leptin and melanocortin pathways in controlling food intake and energy expenditure. These pathways have been individualized within a redundant system of food intake control. Their

---

[1] Inserm Nnutriomique U755, 75004 Paris, France; Université Pierre et Marie Curie – Paris 6, IFR58, 75004 Paris, France; CHRU Pitié Salpétrière, service de Nutrition, Hôtel-Dieu 75004 Paris, France

Conn et al.
Insights into Receptor Function
and New Drug Development Targets
© Springer-Verlag Berlin Heidelberg 2006

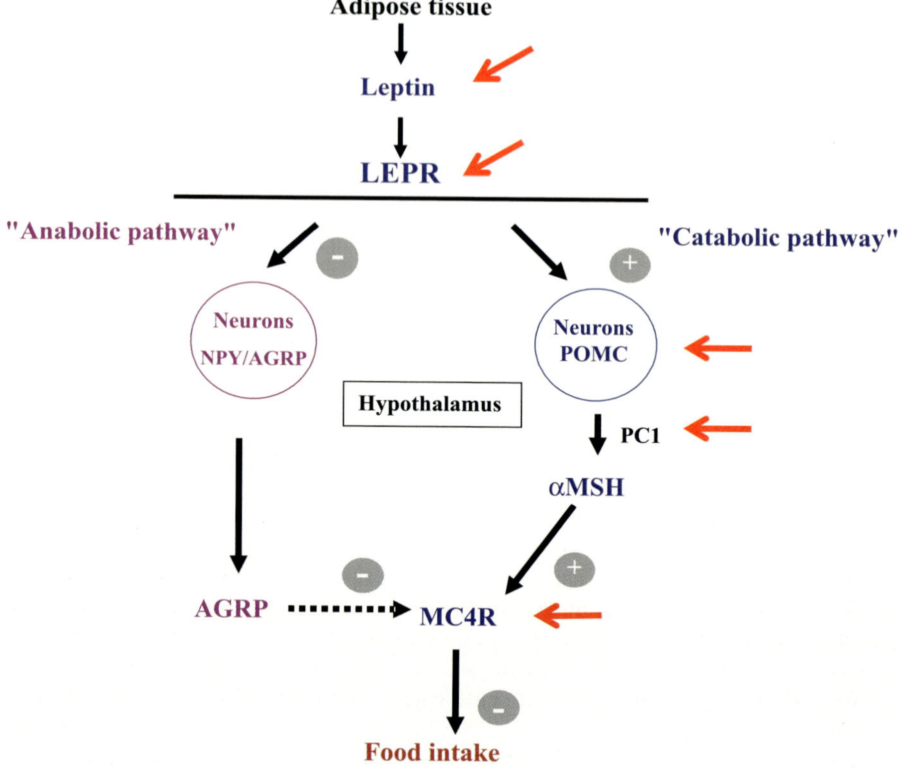

Fig. 1. Human mutations in the leptin/ melanocortin pathway. Schematic representation of leptin and melanocortin pathways. Mutations in leptin, leptin receptor, POMC and PC1 associating severe obesity with multiple endocrine dysfunctions. Lep-R: leptin receptor; POMC: Proopiomelanocortin; αMSH: alpha melanocyte stimulating hormone; AGRP: Agouti Related Protein, MC4R: Melanocortin 4 receptor; PC1: Proconvertase 1. NPY neuropeptide Y →: Location of mutations responsible for monogenic obesity in humans

pivotal role in the control of several endocrine pathways in humans was also elucidated. In addition, research was conducted on the adipose tissue of these human models of extreme obesity.

Leptin and melanocortin pathways

Linkage studies, mouse models and spontaneous human mutations, as well as pharmacological studies, have uncovered the primary role of the leptin/melanocortin pathway in regulating energy homeostasis (Coll et al. 2004). Leptin was discovered more than 10 years ago (Houseknecht et al. 1998). This adipocyte-derived hormone has fully satisfied the criteria for an adiposity signal. In healthy animals and humans, circulating concentrations of leptin highly correlate with body fat mass; leptin crosses the blood-brain barrier and interacts with neurons known to decrease food intake and stimulate thermogenesis. The physiologically active isoforms of the leptin receptor (LEPR) are expressed in these neurons receiving the leptin signal from the periphery. Since its discovery, an increasing amount of knowledge has been gained regarding the mode of action of leptin in the brain. Several brain pathways targeted by leptin have been described. To briefly summarize a complex network, leptin activates anorexigenic neurons (such as proopiomelanocortin [POMC]-derived neurons) through a neural network in the hypothalamic nuclei whereas it inhibits orexigenic neurons (such as NPY/AGRP neurons; NPY = Neuropeptide Y AGRP = Agouti-Related peptide).

The POMC protein is a pivotal actor in this pathway, because of its role in transmitting the leptin signal from the periphery in a central homeostatic response. The production of POMC in the central nervous system is stimulated by leptin, and the post-translational process of the protein gives rise to the production of different peptides harboring various functional properties (Seeley et al. 2004). The nature of the POMC-derived peptides depends on the type of endoproteolytic enzyme present in specific brain locations. In anterior pituitary, the presence of proconvertase 1 (PC1) enzyme allows the production of ACTH and β- lipotropin peptides whereas the presence of both PC1 and PC2 in hypothalamus determines the production of α-, β-, γ-MSH and β-endorphins.

A recent and elegant work using a Cre/loc strategy that creates mice with POMC neuron degeneration showed that POMC-ablation mice develop obesity but with a defect in compensatory hyperphagia, in contrast to the phenotype of POMC-KO mice (Xu et al. 2005). This work clearly emphasizes the role of POMC neurons in controlling adiposity.

One of the important actors in the melanocortin pathway is the melanocortin-4 receptor (MC4R). MC4R is a G-protein coupled receptor with seven transmembrane domains (Cone 2000) that is expressed in the hypothalamus. The importance of MC4R

Table 1. Mutations in human obesity affecting the leptin and melanocortin pathways

Gene	Mode of transmission	Obesity	Associated phenotype
Leptin	Recessive	Severe, from the first days of life	Gonadotropic insufficiency
Leptin receptor	Recessive	Severe, from the first days of life	Gonadotropic, thyrotropic and somatotropic insufficiency
Proopiomelanocortin (POMC)	Recessive	Severe, from the first month of life	ACTH insufficiency; ginger hair
Proconvertase 1 (PC 1)	Recessive	Considerable, from the first month of life	Gonadotropic and corticotropic insufficiency hyperproinsilinemia
Melanocortin 4 receptor (MC4R)	Dominant	Early onset, variable severity larga size	No

in controlling weight homeostasis has been demonstrated in animals. Mice with a genetic invalidation of MC4R (Knockout) develop morbid obesity and increased linear growth. Mice heterozygous for this invalidation present intermediate obesity with various degrees of severity. The use of pharmacological agonists of MC4R in rodents reduces food intake, whereas antagonists of this receptor increase it (Huszar et al. 1997). Mutations of the genes for leptin, its receptor, POMC and proconvertase 1 result in situations of severe obesity with complete penetrance and autosomal recessive transmission (Table 1; reviewed in Barsh et al. 2000; Ozata et al. 1999).

Obesity associated with leptin and LEPR deficiency

Families carrying leptin gene mutations (Montague et al. 1997; Strobel et al. 1998; Farooqi et al. 2001) and one family with three patients affected by a LEPR mutation (Clément et al. 1998) have been recognized. Carriers of these mutations have severe early onset obesity and several endocrine anomalies. The weight or body mass index (BMI) curves of the affected patients are characteristic, as illustrated by the BMI evolution of LEPR mutation carriers (Fig. 2). They show an exponential increase in BMI with severe obesity that develops from the first months or years of life. There is an impulsive pattern of eating behavior and food-seeking disorder, similar to what is observed in patients with Prader-Willi syndrome. The evaluation of body composition in LEPR mutant carriers shows a large amount of total body fat mass 50%. Resting energy expenditure was related to the level of corpulence. In patients with a mutation of leptin or its receptor gene, there is complete failure of puberty through hypogonadotrophic hypogonadism and thyrotropic insufficiency of central origin. Insufficient somatotrophic secretion is also observed in patients with a leptin receptor mutation. A high rate of infection associated with a deficiency in T cell number and function was identified (Farooqi and O'Rahilly 2005). In some patients carrying leptin (Ozata et al. 1999) or LEPR mutations (K. Clément, unpublished observation), there is evidence of spontaneous pubertal development. The follow-up of LEPR-deficient sisters also revealed the normalization of thyroid mild dysfunction in adulthood (K. Clément, unpublished observation).

Leptin-deficient children and adults received great benefit from subcutaneous injections of leptin, resulting in weight loss, mainly of fat mass, with a major effect on reducing food intake and improving other dysfunctions including immunity (Farooqi et al. 2002). An interesting, detailed microanalysis of eating behavior of three adult leptin-deficient subjects before and after three months of leptin treatment revealed a reduced overall food consumption, a slower rate of eating and decreased duration of eating of every meal (Ozata et al. 1999; Williamson et al. 2005). The study supported a role for leptin in influencing the motivation to eat before each meal (Williamson et al. 2005). Hormonal and metabolic changes were evaluated before and after leptin treatment (Licinio et al. 2004). Leptin treatment was able to induce features of puberty even in adults, as illustrated by the effect of leptin treatment in one 27-year-old hypogonadic male. In two women, aged 35 and 40, leptin treatment led to regular menstrual periods and hormonal peaks of progesterone evoking a pattern of ovulation. Although cortisol deficiency was not a feature of leptin or leptin-deficient patients, eight months of leptin

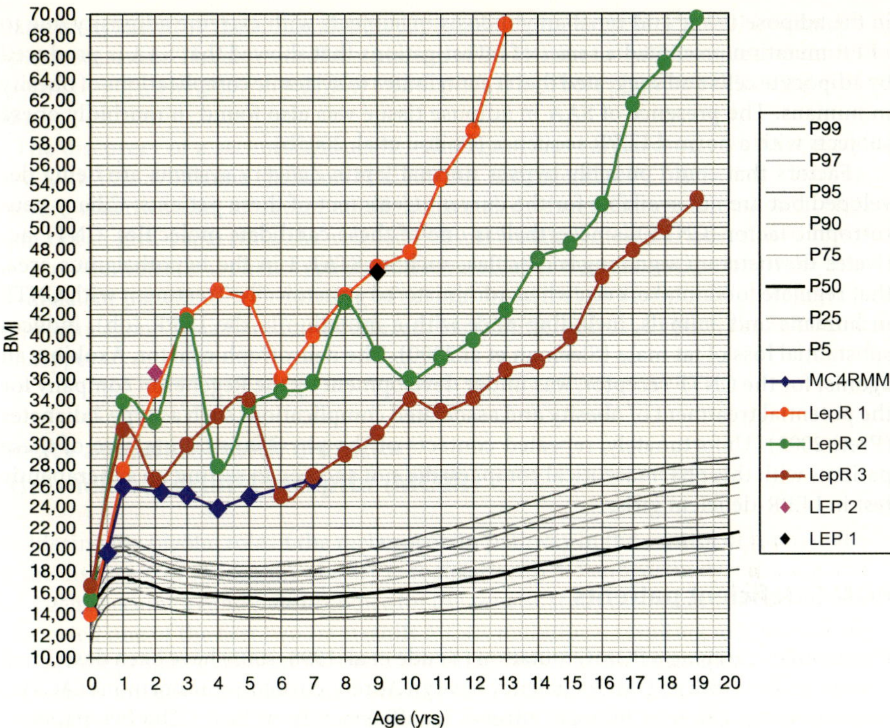

Fig. 2. Body mass index evolution of patients carrying leptin receptor and MC4R mutations at the homozygous state. *Curves* in *bright red*, *dark red* and *green* are BMI curves of patients carrying the leptin receptor mutation; the *curve* in *blue* is the BMI evolution of a patient carrying the MC4R mutation at homozygous state. *Black line* is BMI reference curve for the French populations (50th percentile) *Gray lines* are reference BMI curves. *Upper grey lines* are 99, 97, 95, 90, 75th percentiles. Down grey lines 25 and 5 percentiles) The *pink* and the *black squares* are the BMI measured in the two leptin deficient cousins (at 2 years and 9 year) (Montague et al. 1997)

therapy modified the pulsatility of cortisol, with a greater morning rise of cortisol. Leptin could thus have a previously unsuspected impact on hypothalamic-pituitary-adrenal function in humans. Metabolic parameters of leptin deficient patients also improved with fat mass decrease.

Because of a completely dysfunctional LEPR, any drug could be tested in the lLEPR-deficient sisters. Since 1998, the clinical situation of the two sisters deteriorated, with a maximal weight that reached more than 220 kg at the age of 20. One girl rapidly developed a nephrotic syndrome in 2000 and renal insufficiency, leading to dialysis one year later. Further investigation, including kidney biopsy under polarized light and immunohistochemistry, led to the unambiguous diagnosis of kidney amyloidosis A. Exhaustive clinical, biological and radiological investigations did not detect any common cause of secondary amyloidosis (infection, neoplasia, or inflammatory disease, such as arthritis). Serum amyloid A depots were found in the adipose tissue of both of the LEPR-deficient sisters. The coincidence of Serum amyloid A (SAA) overexpression

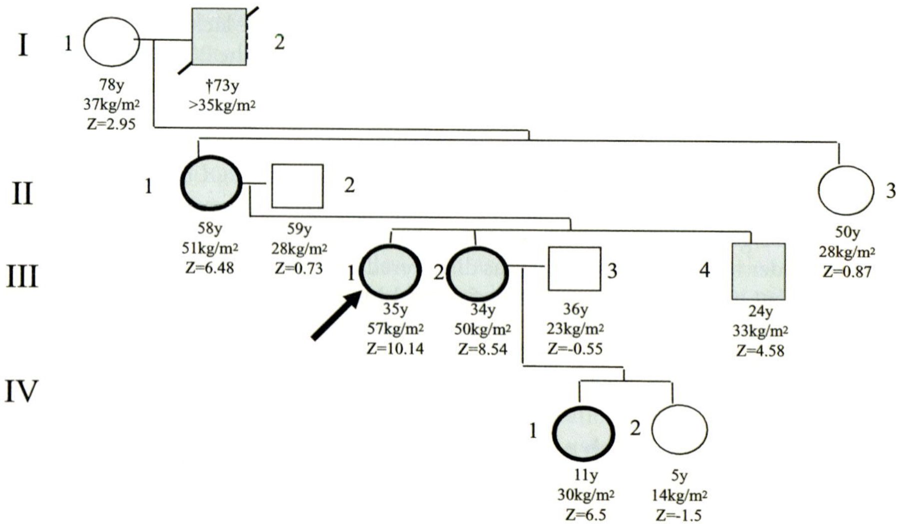

Fig. 3. First MC4R mutation described in a French family. *Filled circles* or *squares* are subjects with the MC4R mutation, which is a frameshift mutation that is transmitted on a dominant basis in three generations. *ZScores* are standard deviation of Body mass index shown in kg/m^2 (from Vaisse et al. 1998). The *arrow* is the index case who was firstly recruited at Hotel-Dieu hospital, Paris, France

animal models (Cre/lox MC4R mice) showing that MC4R neurons in the paraventricular hypothalamus do control food intake but not energy expenditure (Balthasar et al. 2005).

A famous study performed in English children has suggested that bone mineral density and size increase in MC4R mutation carriers (Farooqi et al. 2003). The potential increased bone density observed in MC4R-deficient patients may be caused, at least in part, by a decrease in bone resorption. A decrease of bone resorption markers in the circulation but not of bone formation markers, was found in patients with MC4R homozygous mutations (Elefteriou et al. 2005).

An association between disorders of feeding behavior like "binge eating" and MC4R gene sequence changes has been described (Branson et al. 2003). This finding currently remains very controversial (Farooqi et al. 2003; Gotoda 2003; Herpertz et al. 2003; Hebebrand et al. 2004). Noteworthy in this study, MC4R mutations with functional consequences and common MC4R variants were mixed together. As a general rule, clinical analysis, even if carefully performed, does not easily allow the detection of MC4R mutation-associated obesity, whose phenotype resembles common forms of early onset obesity.

Functional study of MC4R mutations has confirmed the role of these mutations in contributing to obesity development in individual carriers. The response to a melanocortin agonist of mutant MC4R was commonly assayed. After ligand binding, MC4R activation stimulated Gs protein, with a subsequent increase in cAMP levels. The production of intracellular cAMP in response to αMSH demonstrated a broad heterogeneity in the activation of the different MC4R mutants, ranging from normal or partial activation to a total absence of activation (Vaisse et al. 2000).

also
a ro
Nev
cept
nece
of fa

Pol

Ther
com
rare
the g
gene
[(ge
or n
exist
et al
seve
invo
year

a pre
exclu
in di
enco
bette

lepti
for th
chro
in di
centl
have
son a
codii
prom
in co
anoth
and a
mote
gene
genet
For e
untra
Hage

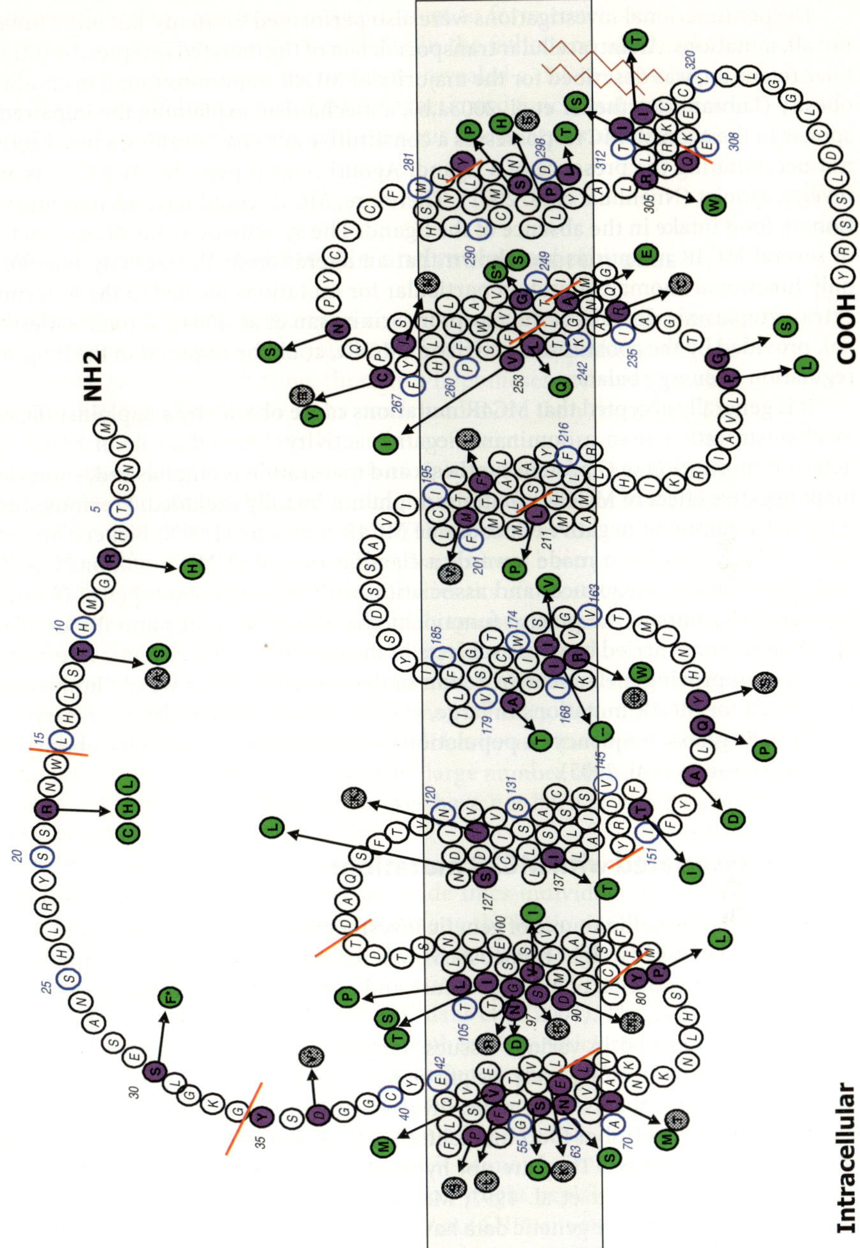

Fig. 4. Examples of MC4R mutations in humans. *Purple circles* are aminoacid position where point mutations were found. *Black arrow* and *green circle* are mutations found with a change of aminoacid. *Red lines* correspond to frameshift mutations

had significantly lower fasting leptin levels compared with subjects who were either heterozygous (AG) or homozygous for the A allele, despite a similar body mass index. More importantly, in 2001, Faroqui et al. studied 13 subjects who were heterozygous for the frameshift mutation delta-G133 of the leptin gene. Their serum leptin concentrations were lower than those in sex- and age-matched controls. Lower leptin levels in these subjects were characterized by an increased prevalence of obesity. In such circumstances, the perception by the central nervous system of the organism's energy reserves could be modified, and/or an inappropriate variation of leptin in response to a variation in energy balance could occur (Farooqi et al. 2001). Polymorphisms of the leptin receptor were also found in obese and diabetic populations, sometimes showing an association but generally a weak one and with inconsistencies (see review in Poitou et al. 2005a).

Similar observations could be made for gene products involved in the melanocortin axis and the role of polymorphisms located in those genes. The POMC gene is located in the human chromosome 2, a region shown to be strongly linked to leptin levels and, to a lesser extent, to obesity. The linkage was initially found in Mexican American families and was replicated in Caucasian and non-Caucasian populations. POMC was considered to be a strong positional candidate, and direct gene screening revealed several polymorphisms, generally of low frequency, located in the coding and non-coding region (review in Carroll et al. 2005). In Mexican Americans, three variants located in the 3′ and 5′ untranslated UTR region, with no functional impact on the protein, and haplotypes revealed association with leptin levels (Hixson et al. 1999). The reason why these variants have been associated with leptin levels is unknown, and another variation located in the POMC gene and/or another nearby gene might be involved. How genetic variation in POMC (or another gene) could contribute to leptin serum level variability is still to be discovered. Several other variants of the POMC gene were found in German (Hinney et al. 1998), Danish (Echwald et al. 1999), Swedish (Suviolahti et al. 2003), Italian (Miraglia del Giudice et al. 2001; Buono et al. 2005), and English (Challis et al. 2002) obese children and young adults and in French diabetic and obese subjects (Delplanque et al. 2000). These variations included base insertion or deletion, missense or silent mutations and are generally uncommon. None of the variants was located in the POMC region encoding for αMSH. Most of these polymorphisms were not associated with obesity phenotypes, including leptin levels, or weight variation in obese subjects, and no functional influence could be predicted except for the Arg236Gly mutation. In obese children from the UK, the frequency of the Arg236Gly mutation was mildly increased, and functional analysis revealed that the mutation prevented the normal processing of γMSH and beta endorphin, resulting in an aberrant fusion protein (Challis et al. 2002). Although able to bind MC4R, this aberrant peptide leads to a decreased activation of the receptor. In general, these studies have been limited in size. In a larger, recent study comprising 1,428 subjects from 248 non-obese families from the UK not specifically recruited for the genetic study of obesity, an association was found between variants spanning the POMC gene (variants located in the 3′ and 5′ UTR) and the waist/hip ratio (WHR) but not with BMI or leptin levels. However, the proportion of the variance in WHR was relatively small 1.1%; Baker et al. 2005). More investigation of the POMC gene in large obese and non-obese population is needed to decipher the role of POMC gene variation in the polygenic nature of common obesity.

Common polymorphisms of MC4R and MC3R were also found both in obese and lean populations. As noted above, a metanalysis and a large screening of the Val103Ile variant of MC4R revealed that this variant is associated with a decreased BMI. Chromosome 20q13, where the MC3R gene is located is a region that has been linked several times with obesity and type 2 diabetes (Perusse et al. 2005), but attempts to associate MC3R common variants and obesity-related phenotypes, including body fat partitioning, have generally been disappointing or, if showing associations, they have not been replicated in independent studies (Li et al. 2000; Schalin-Jantti et al. 2003; Boucher et al. 2002). Explorations of common variants in these candidate genes are underway to further explore their contribution to common forms of human obesity.

Acknowledgements. The programs on genetic research on obesity in French families are supported by the Department of Clinical Research/Assistance Publique Hôpitaux de Paris (Hospital Clinical Research Program), ANR (French National Agency of Research), the INSERM "Avenir" programme, Alfédiam and Aféro.

References

Allison DB, Heo M (1998) Meta-analysis of linkage data under worst-case conditions: a demonstration using the human OB region. Genetics 148:859–865

Baker M, Gaukrodger N, Mayosi BM, Imrie H, Farrall M, Watkins H, Connell JM, Avery PJ, Keavney B (2005) Association between common polymorphisms of the proopiomelanocortin gene and body fat distribution: a family study. Diabetes 54:2492–2496

Balthasar N, Dalgaard LT, Lee CE, Yu J, Funahashi H, Williams T, Ferreira M, Tang V, McGovern RA, Kenny CD, Christiansen LM, Edelstein E, Choi B, Boss O, Aschkenasi C, Zhang CY, Mountjoy K, Kishi T, Elmquist JK, Lowell BB Divergence of melanocortin pathways in the control of food intake and energy expenditure (2005) Cell 123:493–505

Barsh GS, Farooqi IS, O'Rahilly S (2000) Genetics of body-weight regulation. Nature 404:644–651

Biebermann H, Krude H, Elsner A, Chubanov V, Gudermann T, Gruters A (2003) Autosomal-dominant mode of inheritance of a melanocortin-4 receptor mutation in a patient with severe early-onset obesity is due to a dominant-negative effect caused by receptor dimerization. Diabetes 52:2984–2988

Boucher N, Lanouette CM, Larose M, Perusse L, Bouchard C, Chagnon YC (2002) A +2138InsCAGACC polymorphism of the melanocortin receptor 3 gene is associated in human with fat level and partitioning in interaction with body corpulence. Mol Med 8:158–16

Branson R, Potoczna N, Kral JG, Lentes KU, Hoehe MR, Horber FF (2003) Binge eating as a major phenotype of melanocortin 4 receptor gene mutations. New Engl J Med 348:1096–1103

Buono P, Pasanisi F, Nardelli C, Ieno L, Capone S, Liguori R, Finelli C, Oriani G, Contaldo F, Sacchetti L (2005) Six novel mutations in the proopiomelanocortin and melanocortin receptor 4 genes in severely obese adults living in southern Italy. Clin Chem 51:1358–1364

Carroll L, Voisey J, van Daal A (2005) Gene polymorphisms and their effects in the melanocortin system. Peptides 2005 26:1871–1885

Challis BG, Pritchard LE, Creemers JW, Delplanque J, Keogh JM, Luan J, Wareham NJ, Yeo GS, Bhattacharyya S, Froguel P, White A, Farooqi IS, O'Rahilly S (2002) A missense mutation disrupting a dibasic prohormone processing site in pro-opiomelanocortin (POMC) increases susceptibility to early-onset obesity through a novel molecular mechanism. Human Mol Genet 11:1997–2004

Chen AS, Marsh DJ, Trumbauer ME, Frazier EG, Guan XM, Yu H, Rosenblum CI, Vongs A, Feng Y, Cao L, Metzger JM, Strack AM, Camacho RE, Mellin TN, Nunes CN, Min W, Fisher J, Gopal-Truter S, MacIntyre DE, Chen HY, Van der Ploeg LH (2000) Inactivation of the mouse melanocortin-3 receptor results in increased fat mass and reduced lean body mass. Nature Genet 26:97–102

Clement K, Vaisse C, Lahlou N, Cabrol S, Pelloux V, Cassuto D, Gourmelen M, Dina C, Chambaz J, Lacorte JM, Basdevant A, Bougneres P, Lebouc Y, Froguel P, Guy-Grand B (1998) A mutation in the human leptin receptor gene causes obesity and pituitary dysfunction. Nature 392:398–40

Coll AP, Farooqi IS, Challis BG, Yeo GS, O'Rahilly S (2004) Proopiomelanocortin and energy balance: insights from human and murine genetics. J Clin Endocrinol Metab 89:2557–2562

Cone RD (2000) The Melanocortin-4 Receptor. In: Cone RD (ed) The melanocortin receptors. First edn. Humana Press, Totowa, New Jersey, p 405–44

Delplanque J, Barat-Houari M, Dina C, Gallina P, Clement K, Guy-Grand B, Vasseur F, Boutin P, Froguel P (2000) Linkage and association studies between the proopiomelanocortin (POMC) gene and obesity in caucasian families. Diabetologia 43:1554–155

Dempfle A, Hinney A, Heinzel-Gutenbrunner M, Raab M, Geller F, Gudermann T, Schafer H, Hebebrand J (2004) Large quantitative effect of melanocortin-4 receptor gene mutations on body mass index. J Med Genet 41:795–80

Dubern B, Clement K, Pelloux V, Froguel P, Girardet J, Guy-Grand B, Tounian P (2001) Mutational analysis of melanocortin-4 receptor, agouti-related protein, and alpha-melanocyte-stimulating hormone genes in severely obese children. J Pediatr 139:204–209

Echwald SM, Sorensen TI, Andersen T, Tybjaerg-Hansen A, Clausen JO, Pedersen O (1999) Mutational analysis of the proopiomelanocortin gene in Caucasians with early onset obesity. Int J Obes Relat Metab Disord 23:293–29

Elefteriou F, Ahn JD, Takeda S, Starbuck M, Yang X, Liu X, Kondo H, Richards WG, Bannon TW, Noda M, Clement K, Vaisse C, Karsenty G (2005) Leptin regulation of bone resorption by the sympathetic nervous system and CART. Nature 434:514–520

Farooqi IS, O'Rahilly S (2004) Monogenic human obesity syndromes. Recent Prog Horm Res 59:409–42

Farooqi IS, O'Rahilly S (2005) Monogenic obesity in humans. Annu Rev Med 56:443–58

Farooqi IS, Yeo GS, Keogh JM, Aminian S, Jebb SA, Butler G, Cheetham T, O'Rahilly S (2000) Dominant and recessive inheritance of morbid obesity associated with melanocortin 4 receptor deficiency. J Clin Invest 106:271–279

Farooqi IS, Keogh JM, Kamath S, Jones S, Gibson WT, Trussell R, Jebb SA, Lip GY, O'Rahilly S (2001) Partial leptin deficiency and human adiposity. Nature 414:34–35

Farooqi IS, Matarese G, Lord GM, Keogh JM, Lawrence E, Agwu C, Sanna V, Jebb SA, Perna F, Fontana S, Lechler RI, DePaoli AM, O'Rahilly S (2002) Beneficial effects of leptin on obesity, T cell hyporesponsiveness, and neuroendocrine/metabolic dysfunction of human congenital leptin deficiency. J Clin Invest 110:1093–1103

Farooqi IS, Yeo GS, O'Rahilly S (2003) Binge eating as a phenotype of melanocortin 4 receptor gene mutations. New Engl J Med 349:606–609; author reply 606–609

Feitosa MF, Borecki IB, Rich SS, Arnett DK, Sholinsky P, Myers RH, Leppert M, Province MA (2002) Quantitative-trait loci influencing body-mass index reside on chromosomes 7 and 13: the National Heart, Lung, and Blood Institute Family Heart Study. Am J Hum Genet 70:72–82

Gotoda T (2003) Binge eating as a phenotype of melanocortin 4 receptor gene mutations. New Engl J Med 349:606–609; author reply 606–609

Govaerts C, Srinivasan S, Shapiro A, Zhang S, Picard F, Clement K, Lubrano-Berthelier C, Vaisse C (2005) Obesity-associated mutations in the melanocortin 4 receptor provide novel insights into its function. Peptides 26:1909–1919

Hager J, Clement K, Francke S, Dina C, Raison J, Lahlou N, Rich N, Pelloux V, Basdevant A, Guy-Grand B, North M, Froguel P. A polymorphism in the 5′ untranslated region of the human ob gene is associated with low leptin levels. Int J Obes Relat Metab Disord. 1998 22:200–5

Hebebrand J, Friedel S, Schauble N, Geller F, Hinney A (2003) Perspectives: molecular genetic research in human obesity. Obes Rev 4:139–146

Hebebrand J, Geller F, Dempfle A, Heinzel-Gutenbrunner M, Raab M, Gerber G, Wermter AK, Horro FF, Blundell J, Schafer H, Remschmidt H, Herpertz S, Hinney A (2004) Binge-eating episodes are not characteristic of carriers of melanocortin-4 receptor gene mutations. Mol Psychiatry 9:796–800

Heid IM, Vollmert C, Hinney A, Doring A, Geller F, Lowel H, Wichmann HE, Illig T, Hebebrand J, Kronenberg F (2005) Association of the 103I MC4R allele with decreased body mass in 7937 participants of two population based surveys. J Med Genet 42:e21

Heijmans BT, Beem AL, Willemsen G, Posthuma D, Slagboom PE, Boomsma D (2004) Further evidence for a QTL influencing body mass index on chromosome 7p from a genome-wide scan in Dutch families. Twin Res 7:192–196

Herpertz S, Siffert W, Hebebrand J (2003) Binge eating as a phenotype of melanocortin 4 receptor gene mutations. New Engl J Med 349:606–9; author reply 606–609

Hinney A, Becker I, Heibult O, Nottebom K, Schmidt A, Ziegler A, Mayer H, Siegfried W, Blum WF, Remschmidt H, Hebebrand J (1998) Systematic mutation screening of the pro-opiomelanocortin gene: identification of several genetic variants including three different insertions, one nonsense and two missense point mutations in probands of different weight extremes. J Clin Endocrinol Metab 83:3737–3741

Hinney A, Schmidt A, Nottebom K, Heibult O, Becker I, Ziegler A, Gerber G, Sina M, Gorg T, Mayer H, Siegfried W, Fichter M, Remschmidt H, Hebebrand J (1999) Several mutations in the melanocortin-4 receptor gene including a nonsense and a frameshift mutation associated with dominantly inherited obesity in humans. J Clin Endocrinol Metab 84:1483–1486

Hixson JE, Almasy L, Cole S, Birnbaum S, Mitchell BD, Mahaney MC, Stern MP, Mac-Cluer JW, Blangero J, Comuzzie AG (1999) Normal variation in leptin levels in associated with polymorphisms in the proopiomelanocortin gene, POMC. J Clin Endocrinol Metab 84:3187–191

Houseknecht KL, Baile CA, Matteri RL, Spurlock ME (1998) The biology of leptin: a review. J Anim Sci 76:1405–1420

Huszar D, Lynch CA, Fairchild-Huntress V, Dunmore JH, Fang Q, Berkemeier LR, Gu W, Kesterson RA, Boston BA, Cone RD, Smith FJ, Campfield LA, Burn P, Lee F (1997) Targeted disruption of the melanocortin-4 receptor results in obesity in mice. Cell 88:131–141

Jackson RS, Creemers JW, Ohagi S, Raffin-Sanson ML, Sanders L, Montague CT, Hutton JC, O'Rahilly S (1997) Obesity and impaired prohormone processing associated with mutations in the human prohormone convertase 1 gene. Nature Genet 16:303–306

Jackson RS, Creemers JW, Farooqi IS, Raffin-Sanson ML, Varro A, Dockray GJ, Holst JJ, Brubaker PL, Corvol P, Polonsky KS, Ostrega D, Becker KL, Bertagna X, Hutton JC, White A, Dattani MT, Hussain K, Middleton SJ, Nicole TM, Milla PJ, Lindley KJ, O'Rahilly S (2003) Small-intestinal dysfunction accompanies the complex endocrinopathy of human proprotein convertase 1 deficiency. J Clin Invest 112:1550–1560

Jacobson P, Ukkola O, Rankinen T, Snyder EE, Leon AS, Rao DC, Skinner JS, Wilmore JH, Lonn L, Cowan GS, Jr., Sjostrom L, Bouchard C (2002) Melanocortin 4 receptor sequence variations are seldom a cause of human obesity: the Swedish Obese Subjects, the HERITAGE Family Study, and a Memphis cohort. J Clin Endocrinol Metab 87:4442–4446

Jiang Y, Wilk JB, Borecki I, Williamson S, DeStefano AL, Xu G, Liu J, Ellison RC, Province M, Myers RH (2004) Common variants in the 5′ region of the leptin gene are associated with body mass index in men from the National Heart, Lung, and Blood Institute Family Heart Study. Am J Human Genet 75:220–230

Krude H, Biebermann H, Luck W, Horn R, Brabant G, Gruters A (1998) Severe early-onset obesity, adrenal insufficiency and red hair pigmentation caused by POMC mutations in humans. Nature Genet 19:155–157

Krude H, Biebermann H, Schnabel D, Tansek MZ, Theunissen P, Mullis PE, Gruters A (2003) Obesity due to proopiomelanocortin deficiency: three new cases and treatment trials with thyroid hormone and ACTH4–10. J Clin Endocrinol Metab 88:4633–4640

Larsen LH, Echwald SM, Sorensen TI, Andersen T, Wulff BS, Pedersen O (2004) Prevalence of mutations and functional analyses of melanocortin 4 receptor variants identified among 750 men with juvenile-onset obesity. J Clin Endocrinol Metab 90:219–224

Li WD, Joo EJ, Furlong EB, Galvin M, Abel K, Bell CJ, Price RA (2000) Melanocortin 3 receptor (MC3R) gene variants in extremely obese women. Int J Obes Relat Metab Disord 24:206–210

Licinio J, Caglayan S, Ozata M, Yildiz BO, de Miranda PB, O'Kirwan F, Whitby R, Liang L, Cohen P, Bhasin S, Krauss RM, Veldhuis JD, Wagner AJ, DePaoli AM, McCann SM, Wong ML (2004) Phenotypic effects of leptin replacement on morbid obesity, diabetes mellitus, hypogonadism, and behavior in leptin-deficient adults. Proc Natl Acad Sci USA 101:4531–4536

Lubrano-Berthelier C, Cavazos M, Dubern B, Shapiro A, Stunff CL, Zhang S, Picart F, Govaerts C, Froguel P, Bougneres P, Clement K, Vaisse C (2003a) Molecular genetics of human obesity-associated MC4R mutations. Ann NY Acad Sci 994:49–57

Lubrano-Berthelier C, Durand E, Dubern B, Shapiro A, Dazin P, Weill J, Ferron C, Froguel P, Vaisse C (2003b) Intracellular retention is a common characteristic of childhood obesity-associated MC4R mutations. Human Mol Genet 12:145–153

Lubrano-Berthelier C, Le Stunff C, Bougneres P, Vaisse C (2004) A homozygous null mutation delineates the role of the melanocortin-4 receptor in humans. J Clin Endocrinol Metab 89:2028–2032

Marsh DJ, Hollopeter G, Huszar D, Laufer R, Yagaloff KA, Fisher SL, Burn P, Palmiter RD (1999) Response of melanocortin-4 receptor-deficient mice to anorectic and orexigenic peptides. Nature Genet 21:119–122

Miraglia Del Giudice E, Cirillo G, Nigro V, Santoro N, D'Urso L, Raimondo P, Cozzolino D, Scafato D, Perrone L (2002) Low frequency of melanocortin-4 receptor (MC4R) mutations in a Mediterranean population with early-onset obesity. Int J Obes Relat Metab Disord 26:647–651

Miraglia del Giudice E, Cirillo G, Santoro N, D'Urso L, Carbone MT, Di Toro R, Perrone L (2001) Molecular screening of the proopiomelanocortin (POMC) gene in Italian obese children: report of three new mutations. Int J Obes Relat Metab Disord 25:61–7

Montague CT, Farooqi IS, Whitehead JP, Soos MA, Rau H, Wareham NJ, Sewter CP, Digby JE, Mohammed SN, Hurst JA, Cheetham CH, Earley AR, Barnett AH, Prins JB, O'Rahilly S (1997) Congenital leptin deficiency is associated with severe early-onset obesity in humans. Nature 387:903–908

Nijenhuis WA, Oosterom J, Adan RA (2001) AgRP(83–132) acts as an inverse agonist on the human-melanocortin-4 receptor. Mol. Endocrinol. 15:164–171

Ozata M, Ozdemir IC, Licinio J (1999) Human leptin deficiency caused by a missense mutation: multiple endocrine defects, decreased sympathetic tone, and immune system dysfunction indicate new targets for leptin action, greater central than peripheral resistance to the effects of leptin, and spontaneous correction of leptin-mediated defects. J Clin Endocrinol Metab 84:3686–3695

Perusse L, Rankinen T, Zuberi A, Chagnon YC, Weisnagel SJ, Argyropoulos G, Walts B, Snyder EE, Bouchard C (2005) The human obesity gene map: the 2004 update. Obes Res 13:381–490

Poitou C, Lacorte JM, Coupaye M, Bertrais S, Bedel JF, Lafon N, Bouillot JL, Galan P, Borson-Chazot F, Basdevant A, Coussieu C, Clement K (2005a) Relationship between single nucleotide polymorphisms in leptin, IL6 and adiponectin genes and their circulating product in morbidly obese subjects before and after gastric banding surgery. Obes Surg 15:11–23

Poitou C, Viguerie N, Cancello R, De Matteis R, Cinti S, Stich V, Coussieu C, Gauthier E, Courtine M, Zucker JD, Barsh GS, Saris W, Bruneval P, Basdevant A, Langin D, Clement K (2005b) Serum amyloid A: production by human white adipocyte and regulation by obesity and nutrition. Diabetologia 48:519–528

Preti A (2003) Axokine (Regeneron). IDrugs 6:696–701

Schalin-Jantti C, Valli-Jaakola K, Oksanen L, Martelin E, Laitinen K, Krusius T, Mustajoki P, Heikinheimo M, Kontula K (2003) Melanocortin-3-receptor gene variants in morbid obesity. Int J Obes Relat Metab Disord 27:70–74

Seeley RJ, Drazen DL, Clegg DJ (2004) The critical role of the melanocortin system in the control of energy balance. Annu Rev Nutr 24:133–149

Sleeman MW, Anderson KD, Lambert PD, Yancopoulos GD, Wiegand SJ (2000) The ciliary neurotrophic factor and its receptor, CNTFR alpha. Pharm Acta Helv 74:265–272

Snyder EE, Walts B, Perusse L, Chagnon YC, Weisnagel SJ, Rankinen T, Bouchard C (2004) The human obesity gene map: the 2003 update. Obes Res 12:369–439

Srinivasan S, Lubrano-Berthelier C, Govaerts C, Picard F, Santiago P, Conklin BR, Vaisse C (2004) Constitutive activity of the melanocortin-4 receptor is maintained by its N-terminal domain and plays a role in energy homeostasis in humans. J Clin Invest 114:1158–1164

Strobel A, Issad T, Camoin L, Ozata M, Strosberg AD (1998) A leptin missense mutation associated with hypogonadism and morbid obesity. Nature Genet 18:213–215

Suviolahti E, Ridderstrale M, Almgren P, Klannemark M, Melander O, Carlsson E, Carlsson M, Hedenbro J, Orho-Melander M (2003) Pro-opiomelanocortin gene is associated with serum leptin levels in lean but not in obese individuals. Int J Obes Relat Metab Disord 27:1204–1211

Tao YX (2005) Molecular mechanisms of the neural melanocortin receptor dysfunction in severe early onset obesity. Mol Cell Endocrinol 239:1–14

Vaisse C, Clement K, Guy-Grand B, Froguel P (1998) A frameshift mutation in human MC4R is associated with a dominant form of obesity. Nature Genet 20:113–114

Vaisse C, Clement K, Durand E, Hercberg S, Guy-Grand B, Froguel P (2000) Melanocortin-4 receptor mutations are a frequent and heterogeneous cause of morbid obesity. J Clin Invest 106:253–262

Williamson DA, Ravussin E, Wong ML, Wagner A, Dipaoli A, Caglayan S, Ozata M, Martin C, Walden H, Arnett C, Licinio J (2005) Microanalysis of eating behavior of three leptin deficient adults treated with leptin therapy. Appetite 45:75–80

Xu AW, Kaelin CB, Morton GJ, Ogimoto K, Stanhope K, Graham J, Baskin DG, Havel P, Schwartz MW, Barsh GS (2005) Effects of Hypothalamic Neurodegeneration on Energy Balance. PLoS Biol 2005 Nov 29;3(12):e415 [Epub ahead of print]

Yeo GS, Farooqi IS, Aminian S, Halsall DJ, Stanhope RG, O'Rahilly S (1998) A frameshift mutation in MC4R associated with dominantly inherited human obesity. Nature Genet 20:111–112

cAMP- and cGMP-dependent control of lipolysis and lipid mobilization in humans: putative targets for fat cell management

Max Lafontan[1], Michel Berlan[1], Coralie Sengenes[1], Cédric Moro[1], François Crampes[1], and Jean Galitzky[1]

Summary

The mobilization of triglycerides stored in adipose tissue (AT) plays a major role in supplying non-esterified fatty acids (NEFA) to working muscle. Lipolysis, the hydrolysis of fat cell triacylglycerols by hormone-sensitive lipase (HSL), promotes the release of NEFA and glycerol by fat cells. HSL regulation is under the potent control of cAMP and the cAMP-dependent protein kinase, protein kinase A (PKA). In human fat cells, adenylyl cyclase activity and cAMP production are under the positive control of β_1-, β_2- and, to a lesser extent, β_3-adrenergic-receptor (AR)-dependent stimulation and α_2-AR-mediated inhibition. The links between β- and α_2-ARs and adenylyl cyclase activity operate through the activation of Gs and Gi GTP binding proteins respectively. The adrenergic regulation of lipolysis in vitro is complex because of the heterogeneity of the distribution of α_2-/β-ARs in fat cells. A number of in vitro studies have clearly established that the repertoire and the level of expression of ARs in human fat cells differ largely according to the anatomical location and the extent of AT, the sex, the age as well as genetic determinants of the subjects. Fat cells from visceral deposits exhibit the highest β-adrenergic responsiveness and lowest α_2-adrenergic response. Reduced adrenaline-induced lipid mobilization has been reported in the AT of obese subjects. Physiological stimulation of adipocyte α_2-ARs strongly impairs exercise-induced lipolysis in the subcutaneous AT of obese subjects. It is completely reversed by the local administration of an α_2-AR antagonist. Oral administration of an α_2-AR antagonist (yohimbine or idazoxan) promotes sympathetic nervous system (SNS) activation and increased lipid mobilization.

Atrial natriuretic peptide (ANP) has been shown to stimulate cGMP production and exert potent lipolytic effects in human fat cells in vitro. In addition to PKA, cGMP-dependent protein kinase (cGK-I) is involved in the phosphorylation and activation of HSL in human adipocytes. When administered intravenously or physiologically released during exercise, ANP contributes to lipid mobilization in humans. The interplay between the various pathways is discussed as well as putative pharmacological strategies.

Introduction

Obesity has become an alarming problem in Europe and North America, with the prevalence dramatically accentuated in the younger generations. The economic costs

---

[1] Inserm, U586, Unité de Recherches sur les Obésités, Toulouse, F-31342 France; Université Paul Sabatier, Institut Louis Bugnard IFR31, Toulouse, F-31432 France

Conn et al.
Insights into Receptor Function
and New Drug Development Targets
© Springer-Verlag Berlin Heidelberg 2006

General considerations on lipid mobilization and lipolytic pathways in human fat cells

Lipid mobilization from fat stores is an important part of energy turnover. Triacylglycerols stored in white adipose tissue (WAT) are continuously renewed by lipolysis and re-esterification of NEFAs released by the lipolytic process. Mobilization of triacylglycerols is tightly regulated by hormones and requires the activation of fat cell lipolytic enzymes. Lipolysis, the hydrolysis of fat cell triacyl- and diacylglycerols, was considered to be controlled by the activation of hormone-sensitive lipase (HSL). Very recently, a novel triacylglycerol lipase, termed adipose triglyceride lipase (ATGL), desnutrin of iPLA2ζ was discovered. It specifically initiates triacylglycerol hydrolysis, resulting in diacylglycerols and NEFAs (Jenkins et al. 2004; Villena et al. 2004; Zimmermann et al. 2004; Zechner et al. 2005). ATGL seems to play a predominant role in the control of basal lipolysis. However, it cannot be excluded that other AT lipases hydrolyzing triacylglycerols, such as other members of the patatin family or carboxylesterase 3, could also play a role (Soni et al. 2004; Lake et al. 2005). In human fat cells, a recent study has confirmed that HSL is the major lipase for catecholamine and natriuretic peptide-stimulated lipolysis whereas ATGL mediates the hydrolysis of triacylglycerols during basal lipolysis (Langin et al. 2005). Although HSL has the capacity to hydrolyze monoglycerides in vitro, monoglyceride lipase, which is not under hormonal control, is required to obtain complete hydrolysis of monoglycerides in vivo (Fredrikson et al. 1986). The activation of lipolysis promotes NEFA and glycerol release by fat cells. The deregulation of lipolysis has important physiological and clinical implications, and altered lipolysis could be an element predisposing to obesity. In young subjects, a reduced lipolytic action of catecholamines is an early event in obesity (Bougnères et al. 1997). Interindividual variations in AT lipolysis are important for the rate of weight loss. Excessive lipolytic rates, in conjunction with muscle and liver uptake of NEFAs that are not oxidized, may be a major contributor to the metabolic abnormalities found in individuals with visceral or upper body obesity (Jensen 1991, 1997; Lafontan and Berlan 2003). Elevated plasma NEFA concentrations seen in obesity and insulin resistance may be partly caused by impaired insulin-dependent inhibition of intracellular lipolysis or functional alterations of the lipolytic and other antilipolytic pathways controlling fat cell function. Because of the link between elevated circulating NEFA levels and the development of insulin resistance and the metabolic syndrome, AT lipolysis and the hormonal pathways leading to the regulation of lipolysis constitute targets for the pharmaceutical industry.

In man, the major hormones acutely controlling the lipolytic function are catecholamines (i.e., epinephrine and norepinephrine) and insulin (inhibition of lipolysis). Several other factors produced or not by adipocytes, such as prostaglandins, adenosine, neuropeptideY, ketone bodies or peptideYY, also inhibit lipolysis, although the physiological relevance of their in vitro action is still poorly understood. Recently, atrial natriuretic peptide (ANP) has been shown to exert potent lipolytic effects in vitro and to possess lipid mobilizing properties when administered intravenously (Lafontan et al. 2005). WAT is innervated by the sympathetic nervous system (SNS) but human AT innervation is not well understood. Rodent studies have revealed differences in the origins of the sympathetic outflow to WAT, as well as functional differences in the origins of the WAT SNS innervation, that could contribute to the differential propensity for fat cell proliferation and metabolism across WAT depots in vivo (Bowers et al. 2003; Romijn and Fliers 2005). Due to technical limitations, the question has never been investigated in humans; it is an area that merits further investigations. SNS activation is controlled by metabolic and non-metabolic factors (Landsberg 1989). It is considered that a low activity of the SNS is associated with the development of obesity in rodents and humans. No uniform abnormalities in SNS and adrenal medulla activity have been found across all animal models of obesity or all groups of obese human subjects (Bray et al. 1989). Alterations in SNS signalling at the target fat cell level could have a major influence on metabolic events controlled by the SNS, especially lipolysis and thermogenesis.

Lipolysis is a highly regulated process that usually provides adjusted amounts of NEFAs. The delineation of the molecular details of the lipolytic reaction and of the properties of the rate-limiting enzymes of lipolysis (i.e., HSL, MGL and ATGL) has advanced noticeably during the last decade (Langin and Lafontan 2004; Langin et al. 2005; Zechner et al. 2005). In humans, alterations of HSL expression are associated with changes in lipolysis in various physiological and pathological states. Genetic studies have shown that the HSL gene may participate in the polygenic background of obesity (Large et al. 1999). HSL activity is under the potent control of cAMP and cGMP. cAMP-dependent protein kinase (protein kinase A) (PKA) and cGMP-dependent protein kinase of the cGK-1 type (PKB) are involved in the phosphorylation of perilipin and HSL. This enzyme is highly activated by the phosphorylation of Ser^{659} and Ser^{660} residues (Holm et al. 2000) whereas AMP-activated protein kinase phosphorylates Ser^{565} and decreases β-adrenergic receptor-stimulated lipolysis in rodent adipocytes (Sullivan et al. 1994; Daval et al. 2005). Perilipins belong to a family of hydrophobic lipid droplet-associated phosphoproteins. They prevent lipolysis under basal conditions and lose their blocking capacities when phosphorylated by PKA and PKG. Access of activated HSL to the lipid droplet occurs after perilipin phosphorylation and translocation (Tansey et al. 2004; Fig. 1). In addition to perilipins, another family of proteins contributes to the lipolytic complex. Cytosolic lipid-binding proteins [i.e., Adipocyte Lipid Binding Proteins (ALBP)] interact with HSL, favoring its translocation from the cytosol to lipid droplet. ALBP are low molecular weight proteins (15 kDa) that form complexes with fatty acids, retinoids and hydrophobic ligands. They sequester fatty acids inside the cytosol and facilitate lipid and NEFA. Adipocyte and epithelial isoforms of ALBP expressed in AT physically associate with HSL with high affinity and specificity in a fatty acid-dependent manner (Ribarik Coe and Bernlohr 1998; Shen et al. 1999).

Adrenergic control of cAMP production and lipolysis

Activation of the SNS and adrenal medulla function is well known to play a major role in the control of lipid mobilization in rodents and humans. Catecholamines are important stimulators of NEFA release under conditions of stress and during exercise. In vitro studies have revealed that catecholamine action on the human fat cell is complex. Epinephrine and norepinephrine stimulate and/or inhibit lipolysis depending on their relative affinity for the β- and $α_2$-adrenergic receptor subtypes, the relative number of fat cell β and $α_2$-adrenergic receptors expressed in the fat cell and their coupling efficiency to heterotrimeric G-proteins involved in the transduction of the signal (Gs- and

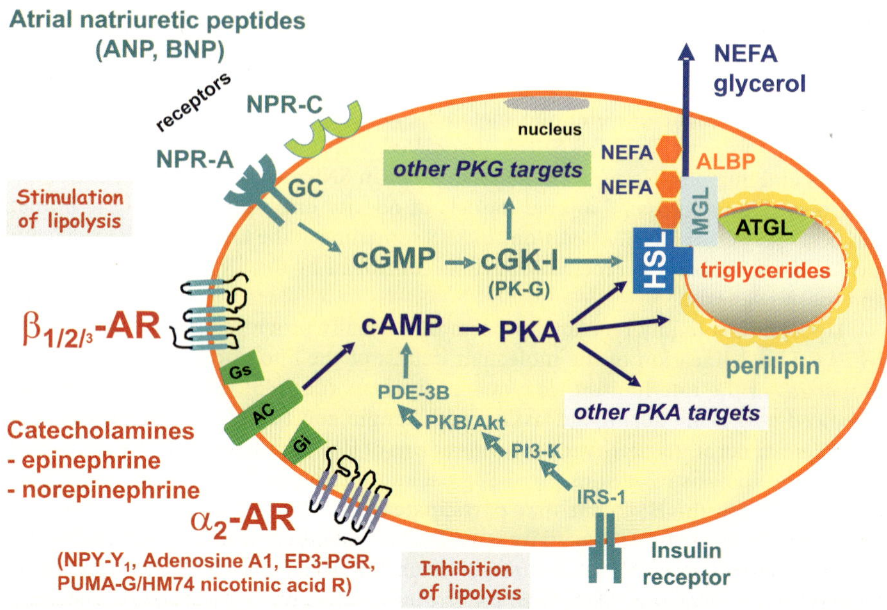

Fig. 1. Control of human fat cell lipolysis: signal transduction pathways for catecholamines via β- and α$_2$-adrenergic receptors (AR), atrial natriuretic peptide via type A receptor (NPR-A) and insulin. AC, adenylyl cyclase; ALBP, adipocyte lipid binding protein; AR, adrenergic receptor; FA, fatty acid; GC, guanylyl cyclase; Gi, inhibitory GTP-binding protein; Gs, stimulatory GTP-binding protein; HSL, hormone-sensitive lipase; ATGL, adipose triglyceride lipase; MGL, monoglyceride lipase; IRS-1, insulin receptor substrate; NEFA, non-esterified fatty acid; PDE-3B, phosphodiesterase 3B; PI3-K, phosphatidylinositol-3-phosphate kinase; PKA, protein kinase A; PKB, protein kinase B/Akt; PKG (cGK-I), protein kinase G. Catecholamines, insulin and various inhibitory receptors negatively coupled to adenylyl cyclase (adenosine, prostaglandins, neuropeptide Y/Peptide YY and nicotinic acid) control cAMP production whereas atrial and brain natriuretic peptides (ANP and BNP) control cGMP production. cAMP and cGMP both contribute to the protein-kinase (PKA and PKG (cGK-I))-dependent phosphorylation of HSL and perilipin. Perilipin phosphorylation induces an important physical alteration of the droplet surface that facilitates the action of HSL on triglyceride hydrolysis. HSL phosphorylation promotes its translocation from the cytosol to the surface of the lipid droplet. Docking of ALBP to HSL favors the efflux of NEFA released by the hydrolysis of triglycerides. PKA and PKG (cGK-I) phosphorylate a number of other substrates (enzymes and transcription factors) that are not shown in the diagram and can also influence the secretion of various adipocyte productions. Insulin receptor stimulation counteracts cAMP production and is without effect on cGMP production (adapted from Lafontan et al. 2005)

Gi-protein, respectively; Fig. 1). Human adipocyte responsiveness to catecholamines differs according to the anatomical location of the AT (Richelsen 1986; Mauriège et al. 1987; Wahrenberg et al. 1989). Decreased catecholamine-induced lipolysis has been reported in the adipocytes of obese men and women (Mauriège et al. 1991,1995a,b; Reynisdottir et al. 1994). Resistance to catecholamine-induced lipolysis in subcutaneous AT is attributed to decreased expression of the lipolytic β$_2$-adrenergic receptor and increased expression of α$_2$-adrenergic receptors and enhanced α$_2$-adrenergic responsiveness (Fig. 2a). Functional differences could also be related to various other downstream elements of the lipolytic cascade (i.e., level of expression of Gs/Gi proteins, modifications of the catalytic and regulatory components of the protein-kinase A complex or the expression level of HSL; Lafontan and Berlan 1993; Lafontan et al. 1997). Decreased catecholamine-induced lipolysis and low HSL expression constitute a possibly primary defect in obesity (Large et al. 1999; Langin et al. 2005). In addition, β-adrenergic receptor signalling alters the expression of adipocyte-specific gene products such as leptin, adiponectin and resistin. Many transcriptional responses to PKA activation are mediated by PKA phosphorylation at Ser^{133} of the cAMP response element binding protein (CREB), which is almost a nuclear protein (Mayr and Montminy 2001).

In the case of β-adrenergic responsiveness, in vitro assays have clearly shown that human fat cell lipolysis is essentially regulated by β_2- and β_1-adrenergic receptor stimulation (Mauriège et al. 1988). No evidence could be provided for a β_3-adrenergic receptor-dependent lipolysis with isoprenaline, norepinephrine or epinephrine stimulation of human fat cells in vitro (Mauriège et al. 1988; Tavernier et al. 1996). It is now well established that CGP12177, the drug usually used to assess a β_3-adrenergic effect in human fat cells, is in fact acting via an atypical state of the β_1-adrenoceptor when this compound is used at higher concentrations to stimulate fat cells. The atypical state of the β_1-adrenergic receptor contributes to the mediation of stimulatory effects induced by commonly used non-conventional partial agonists such as $(-)$ CGP12177 (Konkar

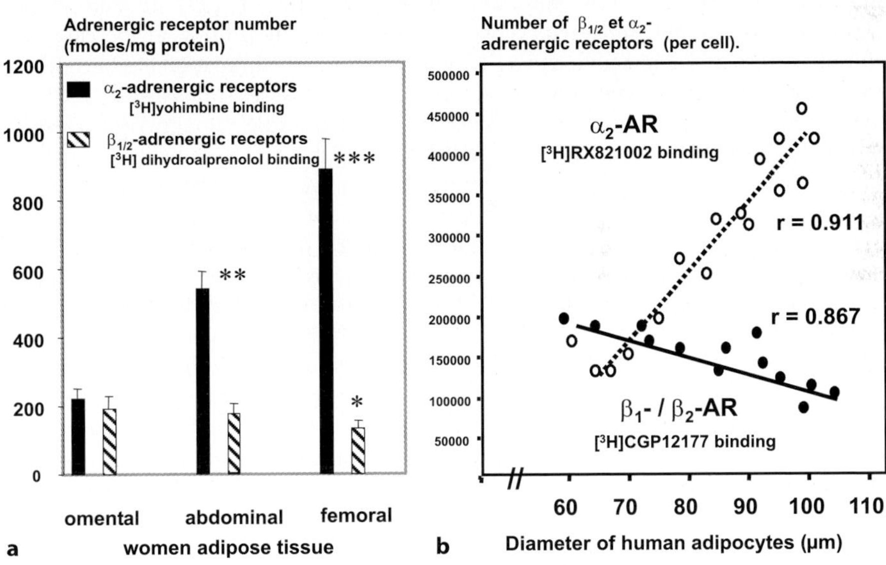

Fig. 2. Expression of $\beta_1 - \beta_2$ and α_2-adrenergic receptors in human fat cells. Radioligands used for $\beta_1 - \beta_2$ and α_2-adrenergic receptor identification are depicted in the figure. (**a**) Expression of adrenergic receptors in fat cells from different AT locations in women. Values are means $\pm$ SEM. ** $P < 0.05$; *** $P < 0.01$; *** $P < 0.001$, vs omental values. (**b**) Relationships existing between $\beta_1-\beta_2$ and α_2-adrenergic receptor expression and mean fat cell diameter in fat cells originating from omental and subcutaneous fat deposits. Binding studies were performed on intact cells with the corresponding ligands

et al. 2000; Kaumann et al. 2001). Confirmation of the lack of a β_3-adrenergic effect in humans has also been provided by in vivo studies. During isoproterenol infusion at doses of 200 ng/kg/min, there was no evidence of a β_3-adrenergic receptor-mediated increase in human lipid mobilization, energy expenditure or lipid oxidation (Schiffelers et al. 2000). Similar conclusions were reached when using in situ microdialysis to delineate the various β-adrenergic receptor subtypes involved in the control of lipolysis in subcutaneous AT (Barbe et al. 1996). Recent genetic approaches have produced a number of controversial results (not all of which can be listed here) concerning the existence of a polymorphism in codon 64 (Trp64Arg) of the β_3-adrenergic receptor gene and the development of obesity and obesity-related disorders (Allison et al. 1998). Subjects bearing the β_3-adrenergic receptor variant, even the heterozygotes, had a lower resting activity of the autonomic nervous system, a dysfunction that could be responsible for the lower resting metabolic rates described in some obese patients (Shihara et al. 1999). To conclude, despite some of the reports on isolated fat cells (Lönnqvist et al. 1995), the studies performed until now have been unable to demonstrate convincingly that, under physiological conditions, the human fat cell β_3-adrenergic receptor contributes to the regulation of lipid mobilization, energy expenditure or lipid oxidation.

Studies performed on adults with long-standing obesity suggest there is a reduced lipolytic sensitivity to catecholamines in subcutaneous abdominal AT. Profound unresponsiveness of the subcutaneous AT to neurally stimulated lipolysis has been described in obese subjects (Dodt et al. 2000). Reduced β_2-adrenergic lipolytic responsiveness has been reported in fat cells from obese subjects (Reynisdottir et al. 1994) or subjects with a reduced isoproterenol sensitivity (Lönnqvist et al. 1992). In addition, an increased antilipolytic responsiveness linked to α_2-adrenergic receptor stimulation has been found in subcutaneous adipocytes from obese subjects of both sexes (Mauriège et al. 1995), and the lipolytic defects have been confirmed in in vivo studies (Bougnères et al. 1997; Carel et al. 1999; Horowitz and Klein 2000b). Using in situ microdialysis, a specific impairment in the capacity of β_2-adrenergic receptor agonists to promote lipolysis has been reported in the subcutaneous abdominal AT of obese adolescent girls (Enoksson et al. 2000). Moreover, when carrying out i.v. administration of selective β_1- and β_2-adrenergic receptor-agonists, the increase in lipolysis and thermogenesis promoted by a selective β_2-adrenergic stimulation (with salbutamol) was reduced in obese subjects. Conversely, β_1-adrenergic receptor-mediated (use of dobutamine) metabolic processes (i.e., lipolysis, thermogenesis and lipid oxidation) were similar in obese and lean men. In conclusion, β_2-adrenergic-mediated increases in thermogenesis and lipid oxidation are impaired in the obese state. It is suspected that a dysfunction of the β_2-adrenergic pathway or of the β_2-adrenergic receptor density may play a role in the etiology or maintenance of a relatively increased fat mass and consequent obesity (Schiffelers et al. 2001). However, it is unknown whether, in parallel, β_2-adrenergic receptor density or coupling is reduced in skeletal muscle cells or blood vessels in obese subjects. Radioligand binding studies have demonstrated a dominant expression of the β_2-adrenergic receptor subtype, which could be of muscular or vascular origin, in skeletal muscle homogenates (Ligget et al. 1988). When using a microdialysis approach in human skeletal muscle, it is observed that only the β_2-adrenergic receptor subtype is of importance for the regulation of lipolysis and the control of local blood flow (Hagström-Toft et al. 1998). Thus, a reduction of β_2-adrenergic efficacy, if occurring, might provide an explanation for the reduced response in lipid oxidation and thermogenesis.

Further evidence for a putative role for β_2-adrenergic receptors in the etiology of obesity is also provided by the discovery of three recently described polymorphisms (the Gln27Glu, Arg16Gly variants and two polymorphic sites T $\rightarrow$ C substitution at -47 and T $\rightarrow$ C substitution at -20, located in the 5′-leader cistron) of the β_2-adrenergic receptor gene that are associated with obesity (Large et al. 1997; Ishiyama-Shigemoto et al. 1999; Mori et al. 1999; Yamada et al. 1999). Genetic variation in the β_2-adrenergic receptor gene influences fat deposition and body size in African-Americans and Hispanic-Americans. Interference for the β_2-adrenergic receptor gene in the distribution of visceral AT, but not subcutaneous AT, was proposed (Lange et al. 2005). Increased body weight, body fat and obesity have also been described in men bearing the Gln27Gln β_2-adrenergic receptor phenotype (Meirhaeghe et al. 2000) and, interestingly, physical activity was able to counterbalance the effect of the genetic predisposition to obesity in such subjects. Thus, obese individuals bearing the Gln27Gln β_2-adrenergic receptor phenotype may benefit from physical activity to reduce their weight (Meirhaeghe et al. 1999). Although some discrepancies still persist, probably related to the poor delineation of the obese phenotypes by the various investigators (Hayakawa et al. 2000; Oberkofler et al. 2000), these findings suggest that polymorphisms in the coding and non-coding sequences in the human β_2-adrenergic receptor gene could be of major importance for obesity, energy expenditure and β_2-adrenergic receptor-dependent lipolytic function. A full β-adrenergic activation of human fat cells usually requires synergistic activation of β_1- and β_2-adrenergic receptors (Lafontan 1994). Even if β_1-adrenergic effects are found to be retained in obese subjects, a β_2-adrenergic defect could be sufficient to alter the normal β-adrenergic responsiveness. In addition, as discussed later in this review, in human fat cells, any reduction of the β_2-adrenergic receptor-mediated lipolytic response will disturb the normal functional balance existing between α_2- and β-adrenergic receptor-mediated effects. The reduction of the lipolytic responsiveness initiated by physiological amines is amplified and it is possible to obtain human fat cells that show very weak lipolytic responses. The β_1-adrenoceptor gene contains two nonsynonomous single nucleotide polymorphisms (SNPs), Ser49Gly and Gly389Arg, which are both functional in human cells lines (Mason et al. 1999; Rathz et al. 2002). Polymorphism of the β_1-adrenoceptor gene influences long-term weight gain and the incidence of adult-onset overweight in women (Linné et al. 2005).

In the case of α_2-adrenergic receptors, in vitro assays in isolated human fat cells have clearly shown variations in the function and expression of human fat cell α_2-adrenergic receptors in various physiological and pathological situations (Lafontan and Berlan 1995). In vitro studies have established that the activation of α_2-adrenergic receptors by epinephrine and norepinephrine impairs the β-adrenergic component of catecholamine-induced lipolysis. In human subcutaneous fat cells, where α_2-adrenergic receptors outnumber β-adrenergic receptors, the preferential recruitment of the α_2-adrenergic receptors at the lowest catecholamine concentrations inhibits lipolysis (Mauriège et al. 1987). The strongest α_2-adrenergic effect has been observed in the adipocytes from subcutaneous AT from both men and women, where α_2-adrenergic receptors are particularly expressed in obese subjects (Mauriège et al. 1991, 1995). Moderate weight loss leads to higher adipose cell lipolytic efficiency, which is associated with changes at the receptor level (mainly increased β_2- and decreased α_2-sensitivities; Mauriège et al. 1999). A number of converging in vitro results suggest an important role for fat cell α_2-adrenergic receptors in the control of lipolysis in obese subjects. One of the major determinants of α_2-adrenergic receptor expression is fat cell hypertrophy; there is a positive correlation between human fat cell diameter and the number of α_2-adrenergic receptor binding sites, whereas the inverse relationship is observed for $\beta_1 - \beta_2$-adrenergic receptors (Fig. 2b). The greater the fat cell size, the lower the lipolytic responsiveness. It is a phenomenon that was also observed in fat cells of various animal models (golden hamster, rabbit). Interestingly, the reduction in fat cell size associated with fat loss induced by calorie restriction is associated with a reduction in α_2-adrenergic receptor expression. The mechanisms driving this cell-size related regulation of fat cell adrenergic receptors expression are unknown.

The utilization of the in situ microdialysis technique has enabled a better demonstration of the relative contribution of the β_{1-2}/α_2-adrenergic receptors in the control of lipid mobilization in vivo. It is a suitable method to study the in vivo lipolytic response of AT to pharmacological or endogenous stimulation (Arner et al. 1990; Barbe et al. 1996; Lafontan and Arner 1996). The administration of an α_2-agonist (clonidine) directly into the microdialysis probe has not been fully conclusive in attributing a physiological role to α_2-adrenergic receptors, due to the potent vasoconstriction induced by clonidine (Galitzky et al. 1993; Millet et al. 1998). In the search for more physiological protocols, exercise was selected as a prerequisite to promote a controlled activation of the SNS. Mild- or moderate-intensity exercise [25–65% of maximal oxygen consumption (VO_2max)] is associated with a 5- to 10-fold increase in fat oxidation above resting levels (Horowitz and Klein 2000a). The catecholamine response to exercise increases lipolysis in AT and presumably of intramuscular triacylglycerols in normal weight subjects. Exercise-induced lipolysis is impaired in subcutaneous AT in obese men and women, and the physiological stimulation of adipocyte α_2-adrenergic receptors during exercise contributes to this impairment. The reduction of lipid mobilization was blocked by local administration of an α_2-adrenergic receptor antagonist (Stich et al. 2000). Striking differences were observed depending on the extent of fat deposits and the intensity of exercise. In heavily trained men, it was impossible to reveal any α_2-adrenergic effect in their reduced subcutaneous fat deposits (de Glisezinski et al. 2001). Moreover, sex-related differences in the lipid-mobilizing efficacy of physical exercise have been revealed in obese women. Unlike the results obtained in obese men, in the subcutaneous AT of obese women the α_2-adrenergic receptors appeared less involved in the control of the lipolytic process during a physiological stimulation of the SNS promoted by exercise (50% maximum O_2 uptake, 60 min duration). Moreover, the weak α_2-adrenergic responsiveness observed in abdominal subcutaneous AT of obese women was also reduced after a moderate hypocalorie diet (Stich et al. 2002). Gender differences in the adrenergic regulation of lipid mobilization have been reported in non-obese patients performing short submaximal exercise bouts (75% maximum O_2 uptake, 30 min duration); however, the nature of the differences was not interpreted by the authors (Hellström et al. 1996). In conclusion, the α_2-adrenergic effects in abdominal subcutaneous AT are always weaker in women, compared with men, whatever the extent of the fat deposits. The noticeable difference between the two genders seems to be related to the exercise-induced profile of the SNS and adrenal medulla activation and catecholamine release. Exercise-induced increments of plasma norepinephrine were quite similar in both sexes whereas plasma epinephrine levels were lower in obese or non-obese women while increasing in men. It could be hypothesized that activation of fat cell α_2-adrenergic receptors by exercise mainly occurs with the release of epinephrine concomitantly with that of norepinephrine. It could be proposed that exercise-related activation of the adrenal medulla and epinephrine release represents a counteracting factor for optimal lipid mobilization. There is strong evidence that supports the hypothesis that, in females, the regulation of the SNS is altered such that sympatho-adrenal activation is attenuated or sympatho-adrenal inhibition is augmented. Attenuated stress-induced increases in plasma catecholamines in women suggest that females are less sensitive and/or less responsive to adrenal medullary activation. This suggestion is supported by findings of gender differences in adrenal medullary catecholamine content, release and degradation (Hinojosa-Laborde et al. 1999). It could be of interest to analyze more deeply this aspect of the question by a comparative study of the impact of exercise bouts of variable intensities and duration in normal and obese men and women. With any given individual, the stress promoted by physical activity is generally determined by the type and intensity of the exercise, the state of physical fitness of the subject, and the nutritional status. Genetic predisposition, sex-related parameters and other factors such as age and fat distribution may also contribute to differences. Additional involvement of the antilipolytic α_2-adrenergic receptors in the control of lipolysis has also been demonstrated during hyperinsulinemia (before and during a euglycemic-hyperinsulinemic clamp). Our study confirmed that, in situ, insulin counteracts the β-adrenergic pathway and showed that the inhibitory antilipolytic α_2-adrenergic component of the lipolytic response, activated by epinephrine, is involved in the control of lipid mobilization during hyperinsulinemia in man.

The reduction of lipid mobilization induced by exercise-induced norepinephrine/epinephrine release is completely counteracted by the local administration of an α_2-adrenergic receptor antagonist. The oral administration of such drugs promotes the SNS activation and increased lipid mobilization. Although such a treatment improves the lipolytic response in some fat deposits, its therapeutic interest is highly questionable since it induces a lipid mobilization that is not followed by a parallel enhancement of NEFA utilization by skeletal muscles.

Insulin: a major antilipolytic agent controlling cAMP degradation

Insulin plays an important role in the control of intracellular cAMP levels and NEFA release. Insulin is a major regulator of lipolysis, as it inhibits lipolysis and NEFA efflux and stimulates glucose uptake by fat cells as well as stimulating fat storage (i.e., it increases the rate of re-synthesis of triacylglycerols from NEFAs; the re-esterification effect). The supply of NEFAs from AT to other tissues is rapidly and strongly inhibited by an increase in the plasma insulin concentration. The cellular mechanisms involved in the inhibition of lipolysis by insulin have been identified and described (Smith and Manganiello 1989; Degerman et al. 1990). Insulin-dependent inhibition of lipolysis operates through cyclic nucleotide phosphodiesterase-3B (PDE-3B) stimulation that promotes cAMP hydrolysis, leading to a reduction in intracellular cAMP levels and inhibition of lipolysis (Fig. 1). Acute insulin treatment activates PDE-3B, reduces cAMP levels and quenches β-adrenergic receptor signaling. By contrast, chronic hyperinsulinemia (typical of type 2 diabetes) enhances β-adrenergic receptor-mediated cAMP production (Hupfeld et al. 2003) but inhibits the activation of PKA (Zhang et al. 2005).

In the postprandial situation or when insulin is infused i.v. using the euglycemic hyperinsulinemic clamp technique, lipolysis is rapidly and strikingly suppressed. On the other hand, reduction of plasma insulin levels, either during fasting, physical exercise or even after acute somatostatin administration, leads to a dramatic increase in the lipolytic rate. A number of circulating factors (such as tumor necrosis factor-alpha, interleukins, insulin itself, NEFAs and glycation products) have been shown to influence insulin action at the target cell level and could lead to hyperglycemia and type 2 diabetes when their action is altered (Pirola et al. 2004).

It seems reasonable to propose that the well-known, upper body, obesity-related metabolic disturbances may be linked to regional variations in lipolysis regulation and NEFA production by insulin. It is clear that moderate changes in fasting insulin levels or insulin sensitivity will alter fat cell lipolysis and fasting plasma NEFA concentrations. Striking differences have been found in fat cell responsiveness to insulin related to the adipose depot, modulated by obesity. Insulin-induced suppression of lipolysis and activation of NEFA re-esterification are reduced in omental compared with subcutaneous fat cells (Bolinder et al. 1983). Various functional differences have been identified at the insulin receptor level and its post-receptor signaling cascade (Zierath et al. 1998). Other partners of the insulin signaling cascade, such as PDE-3B and protein-tyrosine phosphatases (PTPase) involved in the dephosphorylation of the insulin receptor, could also contribute to the modulation of insulin action. Endogenous PTPase activity, including PTPase-1B, is increased in omental AT and may contribute to the relative insulin resistance of this fat depot (Wu et al. 2001).

Increases in the baseline systemic flux of NEFAs have been reported in women with upper body obesity. These increases have been partly attributed to a decreased sensitivity to the antilipolytic effect of insulin, independent of the fat cell size, and to increased lipolytic rates associated with subcutaneous fat cell hypertrophy (Jensen et al. 1989). Subcutaneous abdominal adipocytes are more resistant to the antilipolytic effect of insulin than gluteal adipocytes, independently of cell size (Johnson et al. 2001). In vivo results have confirmed the regional heterogeneity of insulin-regulated NEFA release in vitro. Visceral AT is more resistant to insulin's antilipolytic effects than leg and non-splanchnic body fat (Meek et al. 1999). Nevertheless, visceral fat may be a marker for, but not the source of, excess postprandial NEFAs in obesity, since the increased postprandial NEFA release observed in women with upper body obesity and in type 2 diabetics originates from the non-splanchnic upper body fat, not from visceral fat (Nielsen et al. 2003).

Other cAMP-dependent pathways, adenylyl cyclase inhibitors and antilipolytic agents

Tumor necrosis factor-α (TNF-α) exerts lipolytic effects that seem to be less important for rapid regulation of lipolysis (Kawakami et al. 1987; Ryden et al. 2002). The end-point targets of TNF-α action [i.e., Gi protein, perilipin and insulin receptor substrate-1 (IRS-1)] interfere with the regulation of lipolytic pathways. The Gi effect reported in rodent fat cells modulates the inhibitory effects of adenosine on fat cells and influences the basal lipolytic rate. Recent studies have shown a role for the extracellular signal-regulated kinases (ERKs) in the control of lipolysis by TNF-α in human fat cells (Ryden et al. 2002; Zhang et al. 2002). Activation of ERKs stimulates lipolysis by phosphorylating HSL on Ser[600] (Greenberg et al. 2001). Various autacoid agents and hormonal effectors (prostaglandins, adenosine, neuropeptide Y and Peptide YY) are known to negatively control adenylyl cyclase activity and inhibit cAMP production and lipolysis in human fat cells. The effects are mediated by Gi protein-coupled plasma membrane receptors, the stimulation of which inhibits adenylyl cyclase and cAMP production (Fig. 1). The adenylyl cyclase-coupled receptor subtypes (A1-adenosine receptor, NPY-Y$_1$ receptor and EP-3 prostaglandin receptor) have been identified and quantified in human fat cells. Antagonists of such receptors, relieving the inhibition of cAMP production promoted by the endogenous ligands, enhance the lipolytic activity of fat cells. The physiological relevance of all these in vitro investigations is yet to be established. The receptor of nicotinic acid (niacin), a well-known lipid-lowering drug, has recently been discovered. The orphan G protein-coupled receptor "protein up-regulated in macrophages by interferon γ" (mouse PUMA-G, human HM74a), which is highly expressed in AT, is a nicotinic acid receptor that mediates the antilipolytic and lipid-lowering effect of nicotinic acid in vivo (Lorenzen et al. 2001; Tunaru et al. 2003; Wise et al. 2003). Hydroxybutyrate specifically activates PUMA-G/HM74a at concentrations observed in serum during fasting. Like nicotinic acid, hydroxybutyrate inhibits lipolysis in mouse adipocytes in a PUMA-G-dependent manner. It is the first endogenous physiological ligand described for this receptor. If supported by further clinical studies, these findings suggest a homeostatic mechanism for surviving starvation in which hydroxybutyrate negatively regulates its own production, thereby preventing ketoacidosis and promoting optimal use of fat stores (Taggart et al. 2005). Agonists leading to activation of Gi protein-coupled receptors of the adipocytes limit NEFA release and represent putative antihyperlipidemic drugs. In addition, all these antilipolytic agents that act via the inhibition of cAMP production will also exert leptin-secreting effects.

Lipolytic and lipid-mobilizing effects of natriuretic peptides

The original finding that was the impetus for subsequent studies of the metabolic role of natriuretic peptides (NPs) was the discovery of the lipolytic action of NPs [i.e., atrial natriuretic peptide (ANP), brain natriuretic peptide (BNP) and type C natriuretic peptide (CNP)] in isolated human fat cells. NPs exert potent lipolytic effects similar to those induced by the β-adrenergic receptor agonist, isoproterenol. The relative order of lipolytic potency of the peptides is ANP>BNP>>CNP (Fig. 3a). Binding studies performed on human fat cell membranes using [^{125}I]ANP as a radioligand and various peptide competitors have revealed the presence of natriuretic peptide receptors of the A subtype (NPR-A) in human fat cells (Sengenes et al. 2000).

NPs promoted a strong and sustained increase in intracellular cGMP in human fat cells without any change in cAMP levels (Fig. 3b), and this effect is unrelated to the inhibition of the cGMP-inhibitable phosphodiesterase PDE-3B (Sengenes et al. 2000; Moro et al. 2004a). ANP-induced lipolysis is associated with an increase in the serine phosphorylation of HSL in mature human adipocytes as well as in fat cell precursors differentiated into adipocytes (Fig. 3d). The signal transduction pathway stimulated by ANP is strictly connected to an increase in intracellular cGMP concentrations. The non-hydrolysable analogue of cGMP, 8-bromo-cGMP, mimicked the lipolytic effects of ANP. ANP does not stimulate PKA activity and its inhibition by the PKA inhibitor H-89 does not affect ANP-induced lipolysis. ANP-mediated lipolysis does not involve cross-talk between cGMP and PKA. It is PKG (i.e., cGK-I was the unique form identified in human fat cells) that promotes perilipin and HSL phosphorylation and that underlies the ANP-induced lipolysis (Fig. 3d). The cGMP analogue inhibitor of cGK-I, 8-pCPT-cGMPS, inhibited both HSL phosphorylation/activation and lipolysis (Sengenes et al. 2003). This finding in isolated human fat cells confirmed earlier data obtained on a rat cell-free system where the phosphorylation of HSL by PKG was reported (Khoo et al. 1977; Strålfors and Belfrage 1985). ANP does not modulate the phosphorylation of ERK-1/2 and p38 MAP-kinase. These kinases are not involved in the ANP-mediated HSL phosphorylation since MAP-kinase inhibitors did not affect the ANP-induced HSL phosphorylation.

The lipolytic effect of NPs is completely independent from the major antilipolytic hormone, insulin. Insulin treatment of human fat cells has no effect on the ANP-induced lipolytic response (Sengenes et al. 2000; Moro et al. 2004b, 2005). Since the antilipolytic effects of insulin are mediated by the reduction of intracellular cAMP levels initiated by PDE-3B activation, it is understandable why this hormone does not interfere with the cGMP-dependent NP effects. Identification of the mechanisms involved in the down-regulation of the action of NPs merits attention. Chronic stimulation with NPs and pathological conditions associated with the overproduction of NPs could limit fat cell responsiveness by desensitization of NPR-A activity. Homologous desensitization of the ANP-dependent pathway, subsequent to prior exposure of the adipose cells to ANP, has been shown in isolated fat cells in vitro (Moro et al. 2005).

The occurrence of NP-induced lipolysis is specific to primate fat cells. NPs do not stimulate lipolysis in the fat cells of other species including rats, mice, rabbits and dogs. ANP increased basal cGMP-production 300-fold in human fat cells, whereas it was only stimulated three-fold in rat adipocytes. One of the major explanations for such striking species-related differences in fat cell responsiveness to NPs is that adipocytes from species non-responsive to NPs have a very low expression of the biologically active NPR-A and express mainly plasma membrane clearance NPR-C. Stimulation of the small NPR-A population of these cells is not sufficient to promote the increase in

---

Fig. 3. Effect of natriuretic peptides on in vitro lipolysis, cGMP production in situ lipid mobilization (using microdialysis) and HSL phosphorylation. (**a**) Comparison of the lipolytic effect of natriuretic peptides (ANP, BNP and CNP) and isoproterenol (β-adrenergic receptor agonist concentration-response curve is not shown) on human isolated fat cells. Lipolysis is expressed as a percent of maximal isoproterenol effect (taken from Sengenes et al. 2000). (**b**) Stimulation of cGMP production by ANP in human fat cell. The ANP effect is suppressed by a specific inhibitor of guanylyl cyclase activity (LY 83583)(taken from Sengenes et al. 2003). (**c**) Effect of ANP administration (10 µmol/l) in a microdialysis probe on glycerol release in human subcutaneous AT (taken from Sengenes et al. 2000). Ethanol efflux was used to measure local blood flow changes. ANP promotes local increments in glycerol release and AT blood flow. (**d**) Identification protein kinase G (PKG-cGK-1) in human fat cell (comparison with heart and kidney). Phosphorylation of hormone-sensitive lipase by ANP stimulation of human fat cells (adapted from Sengenes et al. 2003)

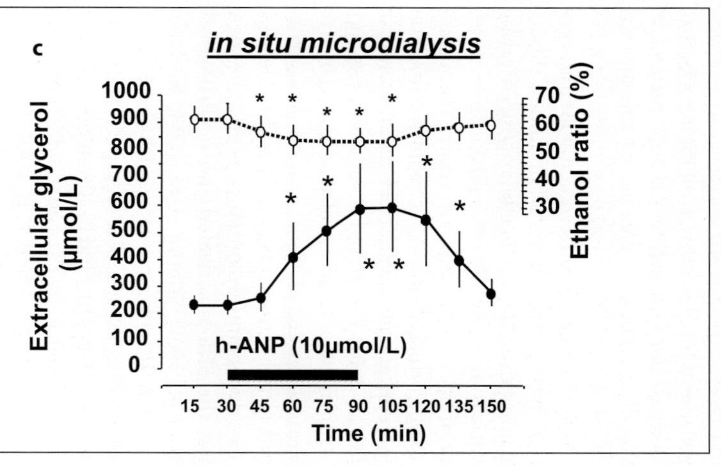
a
in vitro lipolysis
% isoproterenol effect
100 80 60 40 20 0
ANP
BNP
isoproterenol
CNP
0 -11 -10 -9 -8 -7 -6 -5
log [agonists] (mol/L)

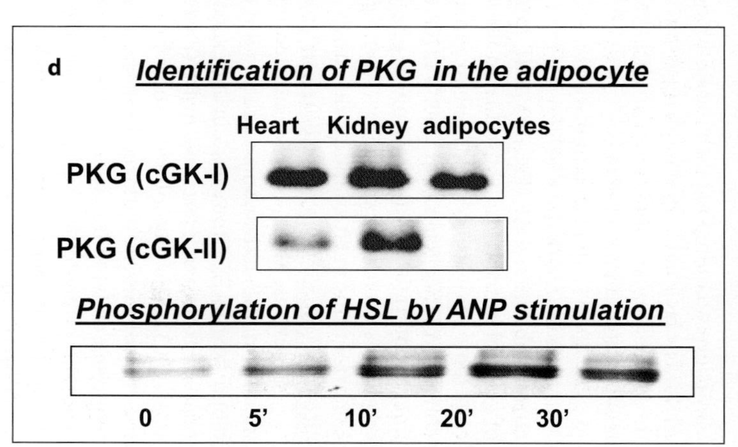
c
in situ microdialysis
Extracellular glycerol (µmol/L)
1000 900 800 700 600 500 400 300 200 100 0
Ethanol ratio (%)
70 60 50 40 30
h-ANP (10µmol/L)
15 30 45 60 75 90 105 120 135 150
Time (min)

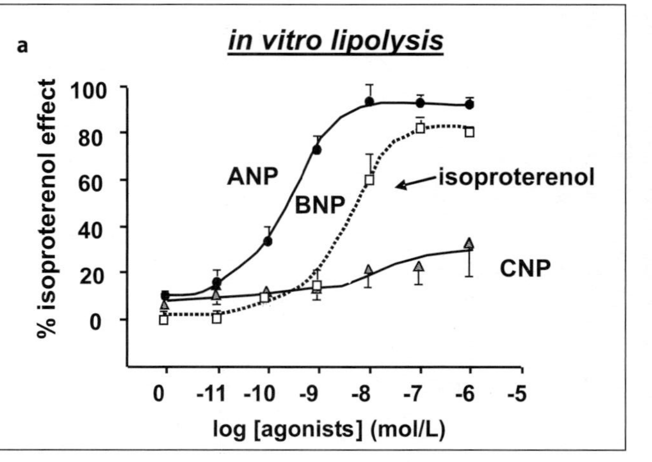
b
cGMP production
pmol/100 mg lipid
400 300 200 100 0
***
Bas   ANP   +LY 83583
0.1 µM   1 µmol/L

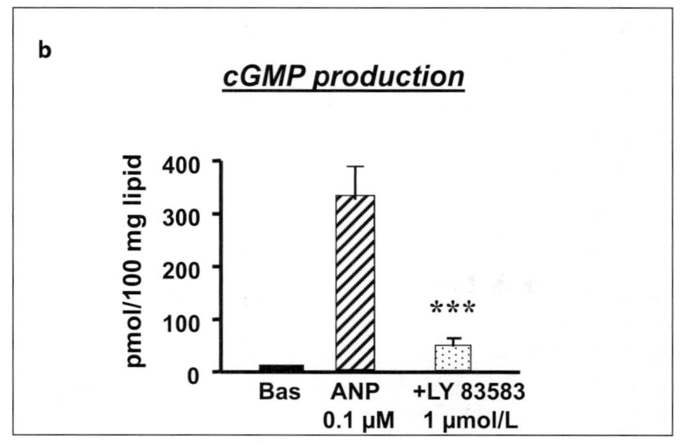
d
Identification of PKG in the adipocyte
Heart   Kidney   adipocytes
PKG (cGK-I)
PKG (cGK-II)
Phosphorylation of HSL by ANP stimulation
0   5'   10'   20'   30'

68   M. Lafontan et al.

cGMP levels required to reach the suitable set-point for HSL activation (Sengenes et al. 2002). The fact that only primates possess this new lipolytic pathway is noteworthy, but the lack of animal models will not help to facilitate future studies on the role and the regulation of the NP system.

The lipolytic action demonstrated in isolated fat cells has been confirmed in vivo with the administration of ANP via a microdialysis probe implanted in human subcutaneous abdominal AT (Fig. 3b). ANP infusion raised the extracellular concentration of glycerol in the AT and vasodilated the vessels draining the fat deposit. Both events contribute to the coordinated enhancement of lipid mobilization in subcutaneous abdominal AT (Sengenes et al. 2000). Intravenous infusion of h-ANP rapidly stimulates lipid mobilization and oxidation at plasma concentrations that are encountered in conditions such as heart failure (Galitzky et al. 2001; Birkenfeld et al. 2005; Fig. 4a). Confirming the in vitro studies, desensitization of the ANP action also occurred *in* vivo after ANP infusion via a microdialysis probe (Moro et al. 2005).

In humans, acute exercise-induced lipid mobilization was considered to depend mainly on SNS activation and the action of catecholamines on fat cells and the local blood flow in AT. However, during exercise, the SNS is activated and ANP/BNP are released from the exercising heart (Siegel et al. 2001; Köning et al. 2003; Niessner et al. 2003). The physiological contribution of ANP, concomitantly with the SNS, to the control exercise-induced lipid mobilization was demonstrated in patients under treatment with β-adrenergic receptor antagonists (Fig. 4b). Oral β-adrenergic receptor blockade promotes strong exercise-related ANP release by the heart, which explains the lipid mobilization continuing under such conditions (Moro et al. 2004a).

Medications that interfere with the natriuretic peptidergic system (i.e., NPR-A agonists and NEP 24.11 inhibitors) could cause metabolic changes. BNP (Nesiritide®) has now been approved for treating acute heart failure since it has beneficial effects on central hemodynamics and the urinary excretion of Na^+ (Colucci et al. 2000; Richards

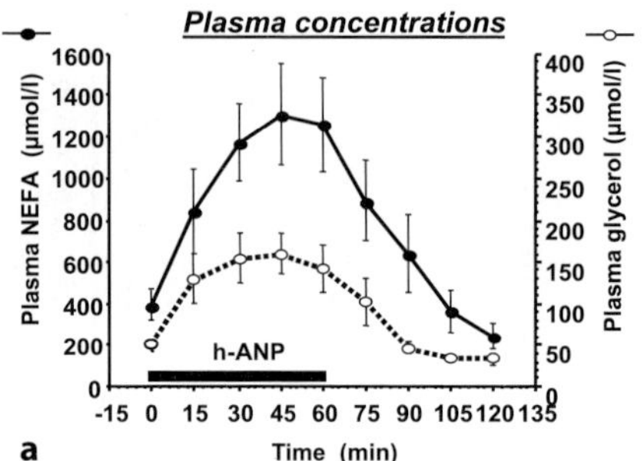

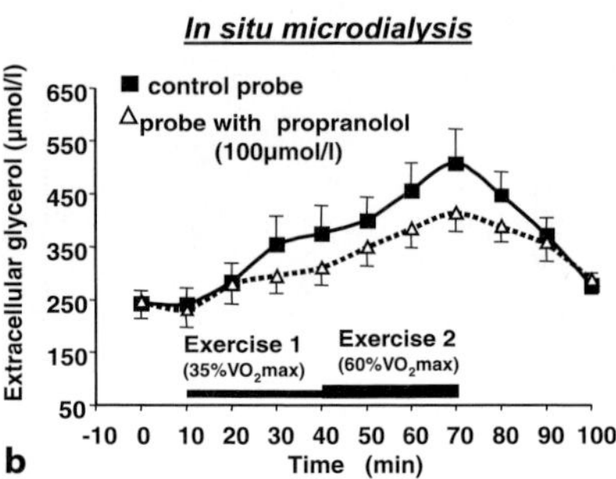

Fig. 4. Lipid mobilization induced by intravenous h-ANP administration or by physical exercise. (a) Effect of intravenous ANP administration (h-ANP at 50 ng/min/kg) on plasma glycerol and NEFA levels in normal weight adult men (adapted from Galitzky et al. 2001). (b) Effect of physical exercise (30 min at 35% and 60% of VO$_2$max, respectively) on lipid mobilization (increase in extracellular glycerol concentration) in human subcutaneous AT using microdialysis. In situ administration of a β-blocking agent (propranolol at 100 µmol/l) only partially blocks exercise-induced lipolysis but completely suppress isoproterenol effect (not shown). Lipid mobilization resistant to β-blockade is due to exercise-induced natriuretic peptides release by the heart and ANP effect on adipocytes (adapted from Moro et al. 2004a)

et al. 2002; Gardner 2003; Maisel 2003). Compared with ANP, BNP seems to be less susceptible to degradation by NEP 24.11 and could be a more potent natriuretic agent than ANP (Abassi et al. 2004). It will be essential to verify if the infusion of BNP, promotes, like an i.v. administration of ANP (Galitzky et al. 2001; Birkenfeld et al. 2005), a potent and sustained lipid mobilizing effect and increases in plasma NEFA levels. This lipid mobilizing effect could alter heart function and counteract some of the beneficial actions of the compound at other tissues. All the drugs known to activate NPR-A/-B-dependent pathways should be evaluated for their metabolic side effects.

Future trends and pharmacological prospects

To conclude, there are only two hormone systems, namely catecholamines and natriuretic peptides, that have acute stimulatory effects on lipolysis and fat mobilization in humans. Both operate through two strictly independent pathways that control cAMP and cGMP production, respectively. A number of alterations in the responsiveness to catecholamines have been described in human fat cells in obese patients. AT (e.g., fat cells and the vascular bed of AT) is a new and unsuspected target organ of NPs. Induction of lipolysis and lipid mobilization must now be included in the numerous physiological actions of NPs. Further fundamental and clinical studies will be required to answer the various questions raised by the discovery of the metabolic effects of NPs. The effect of β-AR antagonist treatment on NP release merits further study. Moreover, it must be determined whether, in addition to the well-known alterations in SNS activity, a chronic and sustained increase in NP production influences cachexia in patients with congestive heart failure or other diseases related to modified NP production. Finally, all the drugs known to modulate NP-dependent pathways should be evaluated for their putative metabolic side effects when given for the management of cardiovascular disease.

Deregulation of lipid metabolism has long been recognized as essential in the development of obesity and the metabolic syndrome. Lipolysis plays a pivotal role in controlling the quantity of triglycerides stored in fat depots and in determining plasma NEFA levels. Critical steps in this catabolic process constitute targets for strategies to fight against obesity and to improve the poor metabolic profile of patients with the metabolic syndrome. First of all, activators of lipolysis could present a pharmacological interest to induce mobilization in fat deposits that are resistant to hormone-induced lipolysis. However, they must be associated with agents that stimulate the oxidation of fatty acids by skeletal muscle and energy expenditure. The lipolytic and thermogenic β_3-adrenergic agonists that are highly efficient in decreasing fat mass and insulin resistance in rodents are not effective in humans. Several potential explanations could be proposed. Unlike in rodents, whose white fat cells express a considerable number of β_3-adrenoceptors, these receptors are expressed at very low levels in human white adipocytes. Also, adult humans have very little brown AT, the tissue specializing in thermogenesis in rodents. Finally, the first generation of β_3-adrenergic agonists had poor selectivity, low efficacies and potencies for the human β_3-adrenergic receptor. Nevertheless, conceptually the approach remains interesting. Ideally, a promising strategy could be to stimulate lipolysis and the utilization of the fatty acids by newly formed brown fat cells (Tiraby and Langin 2003). Concomitant stimulation of fat oxidation in skeletal muscle is an important alternative strategy.

72   M. Lafontan et al.

Ferraro RT, Eckel RH, Larson DE, Fontvielle A-M, Rising R, Jensen DR, Ravussin E (1993) Relationship between skeletal muscle lipoprotein lipase activity and 24-hour macronutrient oxidation. J Clin Invest 92:441–445

Fredrikson G, Tornqvist H, Belfrage P (1986) Hormone-sensitive lipase and monoacylglycerol lipase are both required for complete degradation of adipocyte triacylglycerol. Biochim Biophys Acta 876:288–293

Galitzky J, Taouis M, Berlan M, Rivière D, Garrigues M, Lafontan M (1988) Alpha2-antagonist compounds and lipid mobilization:evidence for a lipid mobilizing effect of oral yohimbine in healthy male volunteers. Eur J Clin Invest 18:587–594

Galitzky J, Vermorel M, Lafontan M, Montastruc P, Berlan M (1991) Thermogenic and lipolytic effect of yohimbine in the dog. Br J Pharmacol 104:514–518

Galitzky J, Lafontan M, Nordenström J, Arner P (1993) Role of vascular alpha$_2$-adrenoceptors in regulating lipid mobilization from human adipose tissue. J Clin Invest 91:1997–2003

Galitzky J, Sengenes C, Thalamas C, Marques M-A, Senard J-M, Lafontan M, Berlan M (2001) The lipid mobilizing effect of atrial natriuretic peptides is unrelated to sympathetic nervous system activation or obesity in young men. J Lipid Res 42:536–544

Gardner DG (2003) Natriuretic peptides:markers or modulators of cardiac hypertrophy. Trends Endocrinol Metab 14:411–416

Greenberg AS, Shen W-J, Muliro K, Souza SC, Roth RA, Kraemer FB (2001) Stimulation of lipolysis and hormone-sensitive lipase via the extracellular signal-regulated kinase pathway. J Biol Chem 276:45456–45461

Grundy SM (2004) Obesity, metabolic syndrome and cardiovascular disease. J Clin Endocrinol Metab 89:2595–2600

Grundy SM, Brewer HBJ, Cleeman JI, Smith SCJ, Lenfant C, National Heart L, and Blood Institute, Association AH (2004) Definition of metabolic syndrome: report of the National Heart, Lung, and Blood Institute/American Heart Association Conference on scientific issues related to definition. Arterioscl ThrombVasc Biol 24:e13–e18

Hagström-Toft E, Enoksson S, Moberg E, Bolinder J, Arner P (1998) β-adrenergic regulation of lipolysis and blood flow in human skeletal muscle in vivo. Am J Physiol 275:E909–E916

Hayakawa T, Nagai Y, Kahara T, Yamashita H, Takamura T, Abe T, Nomura G, Kobayashi K (2000) Gln27Glu and Arg16Gly polymorphisms of the beta2-adrenergic receptor gene are not associated with obesity in japanese men. Metabolism 49:1215–1218

Hellström L, Blaak E, Hagstrom-Toft E (1996) Gender differences in adrenergic regulation of lipid mobilization during exercise. Int J Sports Med 17:439–447

Hinojosa-Laborde C, Chapa I, Lange D, Haywood JR (1999) Gender differences in sympathetic nervous system regulation. Clin Exp Pharmacol Physiol 26:122–126

Holm C, Osterlund T, Laurell H, Contreras JA (2000) Molecular mechanisms regulating hormone-sensitive lipase and lipolysis. Annu Rev Nutr 20:365–393

Horowitz J, Klein S (2000a) Lipid metabolism during endurance exercise. Am J Clin Nutr 72 (Suppl):558S–563S

Horowitz JF, Klein S (2000b) Whole body and abdominal lipolytic sensitivity to epinephrine is suppressed in upper body obese women. Am J Physiol Endocrinol Metab 278:E1144–E1152

Hupfeld CJ, Dalle S, Olefsky JM (2003) Arrestin-1 down-regulation after insulin treatment is associated with supersensitization of β2-adrenergic receptor Gαs signaling in 3T3-L1 adipocytes. Proc Natl Acad Sci USA 100:161–166

Ishiyama-Shigemoto S, Yamada K, Yuan X, Ichikawa F, Nonaka K (1999) Association of polymorphisms in the beta2-adrenergic receptor gene with obesity, hypertriglyceridemia and diabetes mellitus. Diabetologia 42:98–101

Jenkins CM, Mancuso DJ, Yan W, Sims HF, Gibson B, Gross RW (2004) Identification, cloning, expression and purification of three novel calcium-independent phospholipase A2 family members possessing triacylglycerol lipase and acylglycerol transacylase activities. J Biol Chem 279:48968–48975

Jensen M, Haymond M, Rizza R, Cryer P, Miles J (1989) Influence of body fat distribution on free fatty acid metabolism in obesity. J Clin Invest 83:1168–1173

Jensen MD (1991) Regulation of forearm lipolysis in different types of obesity. In vivo evidence for adipocyte heterogeneity. J Clin Invest 87:187–193

Jensen MD (1997) Lipolysis: contribution from regional fat. Annu Rev Nutr 17:127–139

Johnson JA, Fried SK, Pi-Sunyer FX, Albu JB (2001) Impaired insulin action in subcutaneous adipocytes from women with visceral obesity. Am J Physiol 280:E40–E49

Karpe F, Frayn KN (2004) The nicotinic acid receptor-a new mechanism for an old drug. Lancet 363:1892–1894

Kaumann A, Engelhardt S, Hein L, Molenaar P, Lohse M (2001) Abolition of (−)CGP12177-evoked cardiostimulation in double beta1/beta2-adrenoceptor knock out mice. Obligatory role of beta1-adrenoceptors for putative beta4-adrenoceptor pharmacology. Naunyn Schmied Arch Pharmacol 363:87–93

Kawakami M, Murase T, Igawa H, Ishibashi S, Mori N, Takagu F, Shibata S (1987) Human recombinant TNF suppresses lipoprotein lipase activity and stimulates lipolysis in 3T3-L1 cells. J Biochem 101:331–338

Khoo JC, Sperry PJ, Gill GN, Steinberg D (1977) Activation of hormone-sensitive lipase and phosphorylase kinase by purified cyclic GMP-dependent protein kinase. Proc Natl Acad Sci USA 74:4843–4847

Köning D, Schumacher O, Heinrich L, Schmid A, Berg A, Dickhuth HH (2003) Myocardial stress after competitive exercise in professional road cyclists. Med Sci Sports Exerc 35:1679–1683

Konkar AA, Zhai Y, Granneman JG (2000) β1-adrenergic receptors mediate β3-adrenergic-independent effects of CGP12177. Mol Pharmacol 57:252–258

Lafontan M (1994) Differential recruitment and differential regulation by physiological amines of fat cell β1-, β2- and β3-adrenergic receptors expressed in native fat cells and transfected cell lines. Cell Signalling 6:363–392

Lafontan M (2005) Fat cells:Afferent and efferent messages define new approaches to treat obesity. Annu. Rev. Pharmacol Toxicol 45:119–146

Lafontan M, Arner P (1996) Application of in situ microdialysis to measure metabolic and vascular responses in adipose tissue. Trends Pharmacol Sci 17:309–313

Lafontan M, Berlan M (1993) Fat cell adrenergic receptors and the control of white and brown fat cell function. J Lipid Res 34:1057–1091

Lafontan M, Berlan M (1995) Fat cell α_2-adrenoceptors:the regulation of fat cell function and lipolysis. Endocrine Rev 16:716–738

Lafontan M, Berlan M (2003) Do regional differences in adipocyte biology provide new patho-physiological insights? Trends Pharmacol Sci 24:276–283

Lafontan M, Barbe P, Galitzky J, Tavernier G, Langin D, Carpéné C, Bousquet-Melou A, Berlan M (1997) Adrenergic regulation of adipocyte metabolism. Human Reprod 12:6–20

Lafontan M, Moro C, Sengenes C, Galitzky J, Crampes F, Berlan M (2005) An unsuspected metabolic role for atrial natriuretic peptides:The control of lipolysis, lipid mobilization and systemic nonesterified fatty acids levels in humans. Arterioscler Thromb Vasc Biol 25:2032–2042

Lake AC, Sun Y, Li JL, Kim JE, Johnson JW, Li D, Revett T, Shi HH, Liu W, Paulsen JE, Gimeno RE (2005) Expression, regulation and triglyceride hydrolase activity of adiponutrin family members. J Lipid Res 46:2477–2487

Landsberg L (1989) The sympathoadrenal system, obesity and hypertension:an overview. J Neurosci Meth 34:179–186

Lange LA, Norris JM, Langefeld CD, Nicklas BJ, Wagenknecht LE, Saad MF, Bowden DW (2005) Association of adipose tissue deposition and beta-2 adrenergic receptor variants:the IRAS family study. Int J Obesity 29:449–457

Langin D, Lafontan M (2004). Lipolysis and lipid mobilization in human adipose tissue. In: Bray GA, Bouchard C (eds) Handbook of obesity. etiology and pathophysiology. 2nd Edition. Marcel Dekker, Inc, New York. Basel, p 515–532

Langin D, Dicker A, Tavernier G, Hoffstedt J, Mairal A, Ryden M, Arner E, Sicard A, Jenkins CM, Viguerie N, van Harmelen V, Gross RW, Holm C, Arner P (2005) Adipocyte lipases and defect of lipolysis in human obesity. Diabetes 54:3190–3197

Large V, Hellström L, Reynisdottir S, Lönnqvist F, Eriksson P, Lannfelt L, Arner P (1997) Human beta2-adrenoceptor gene polymorphism are highly frequent in obesity and associated with altered adipocyte beta2-adrenoceptor function. J Clin Invest 100:3005–3013

Large V, Reynisdottir S, Langin D, Fredby K, Klannemark M, Holm C, Arner P (1999) Decreased expression and function of adipocyte hormone-sensitive lipase in subcutaneous fat cells of obese subjects. J Lipid Res 40:2059–2065

Ligget SB, Shah SD, Cryer PE (1988) Characterization of β-adrenergic receptors in human skeletal muscle obtained by needle biopsy. Am J Physiol 254:E795–E798

Linné Y, Dahlman I, Hoffstedt J (2005) β1-adrenoceptor gene polymorphism predicts long-term changes in body weight. Int J Obesity 29:458–461

Lönnqvist F, Wahrenberg H, Hellstrom L, Reynisdottir S, Arner P (1992) Lipolytic catecholamine resistance due to decreased β2-adrenoceptor expression in fat cells. J Clin Invest 90:2175–2186

Lönnqvist F, Thörne A, Nilsell K, Hoffstedt J, Arner P (1995) A pathogenic role of visceral fat β3-adrenoceptors in obesity. J Clin Invest 95:1109–1116

Lorenzen A, Stannek C, Lang H, Andrianov V, Kalvinsh I, Schwabe U (2001) Characterization of a G protein-coupled receptor for nicotinic acid. Mol Pharmacol 59:349–357

Lowe DB, Magnuson S, Qi N, Campbell A-M, Cook J, Hong Z, Wang M, Rodriguez M, Achebe F, Kluender H (2004) In vitro SAR of (5-(2H)-isoxazolonyl) ureas, potent inhibitors of hormone-sensitive lipase. Bioorg Medicinal Chem Lett 14:3155–3159

Maisel AS (2003) Nesiritide: a new therapy for the treatment of heart failure. Cardiovasc Toxicol 3:37–42

Mason DA, Moore JD, Green SA, Liggett SB (1999) A gain of function polymorphism in a G-protein coupling domain of the human beta1-adrenergic receptor. J Biol Chem 274:12670–12674

Mauriège P, Galitzky J, Berlan M, Lafontan M (1987) Heterogeneous distribution of beta- and alpha2-adrenoceptor binding sites in human fat cells from various fat deposits: functional consequences. Eur J Clin Invest 17:156–165

Mauriège P, Pergola GD, Berlan M, Lafontan M (1988) Human fat cell beta-adrenergic receptors: beta agonist-dependent lipolytic responses and characterization of beta-adrenergic binding sites on human fat cell membranes with highly selective beta1-antagonists. J Lipid Res 29:587–601

Mauriège P, Després JP, Prud'homme D, Pouliot MC, Marcotte M, Tremblay A, Bouchard C (1991) Regional variation in adipose tissue lipolysis in lean and obese men. J Lipid Res 32:1625–1633

Mauriège P, Marette A, Atgié C, Bouchard C, Thériault G, Bukowiecki LK, Marceau P, Biron S, Nadeau A, Després J-P (1995a) Regional variation in adipose tissue metabolism of severely obese premenopausal women. J Lipid Res 36:672–684

Mauriège P, Prud'homme D, Lemieux S, Tremblay A, Després J-P (1995b) Regional differences in adipose tissue lipolysis from lean and obese women: existence of postreceptor alterations. Am J Physiol 269:E341–E350

Mauriège P, Imbeault P, Langin D, Lacaille M, Alméras N, Tremblay A, Després JP (1999) Regional and gender variations in adipose tissue lipolysis in response to weight loss. J Lipid Res 40:1559–1571

Mayr B, Montminy M (2001) Transcriptional regulation by the phosphorylation-dependent factor CREB. Nature Rev Mol Cell Biol 2:569–609

McGarry JD (2002) Dysregulation of fatty acid metabolism in the etiology of type 2 diabetes. Diabetes 51:7–18

Meek SE, Nair KS, Jensen MD (1999) Insulin regulation of regional free fatty acid metabolism. Diabetes 48:10–14

Meirhaeghe A, Helbecque N, Cottel D, Amouyel P (1999) β2-adrenoceptor gene polymorphism, body weight and physical activity. The Lancet 353:236

Meirhaeghe A, Cottel D, Helbecque N, Amouyel P (2000) Impact of polymorphisms in the human β2-adrenoceptor gene on obesity in French population. Int J Obesity Relat Metab Disord 24:382–387

Millet L, Barbe P, Lafontan M, Berlan M, Galitzky J (1998) Catecholamine effects on lipolysis and blood flow in human abdominal and femoral adipose tissue. J Appl Physiol 85:180–188

Mori Y, Kim-Motoyama H, Ito Y, Katakura T, Yasuda K, Ishiyama-Shigemoto S, Yamada K, Akanuma Y, Ohashi Y, Kimura S, Yazaki Y, Kadowaki T (1999) The Gln27Glu beta2-adrenergic receptor variant is associated with obesity due to subcutaneous fat accumulation. Biochem Biophys Res Commun 258:138–140

Moro C, Crampes F, Sengenes C, de Glisezinski I, Galitzky J, Thalamas C, Lafontan M, Berlan M (2004a) Atrial natriuretic peptide contributes to the physiological control of lipid mobilization in humans. FASEB J 18:908–910

Moro C, Galitzky J, Sengenes C, Crampes F, Lafontan M, Berlan M (2004b) Functional and pharmacological characterization of the natriuretic peptide-dependent lipolytic pathway in human fat cells. J Pharmacol Exp Ther 308:984–992

Moro C, Polak J, Richterova B, Sengenes C, Pelikanova T, Galitzky J, Stich V, Lafontan M, Berlan M (2005) Differential regulation of atrial natriuretic peptide- and adrenergic receptor-dependent lipolytic pathways in human adipose tissue. Metabolism 54:122–131

Nguyen M, Satoh S, Favelyukis S, Babendure JL, Imamura T, Sbodio JI, Zalewsky J, Dahiyat BI, Chi N-W, Olefsky JM (2005) JNK and tumor necrosis factor-a mediate free fatty acid-induced insulin resistance in 3T3-L1 adipocytes. J Biol Chem 280:35361–35371

Nielsen S, Guo Z, Albu JB, Klein S, O'Brien PC, Jensen MD (2003) Energy expenditure, sex, and endogenous fuel availability in humans. J Clin Invest 111:981–988

Niessner A, Ziegler S, Slany J, Billensteiner E, Woloszczuk W, Geyer G (2003) Increases in plasma levels of atrial and brain natriuretic peptides after running a marathon: are their effects partly counterbalanced by adrenocortical steroids? Eur J Endocrinol 149:555–559

Oberkofler H, Esterbauer H, Hell E, Krempler F, Patsch W (2000) The Gln27Glu polymorphism in the beta2-adrenergic receptor gene is not associated with morbid obesity in Austrian women. Int J Obesity Relat Metab Disord 24:388–390

Paolisso G, Tacunni FA, Foley JE, Bogardus C, Howard BV, Ravussin E (1995) A high concentration of fasting plasma non-esterified fatty acids is a risk factor for the development of NIDDM. Diabetologia 38:1213–1217

Paolisso G, Gambardella A, Tagliamonte MR, Saccomano F, Salvatore T, Gualdiero P, D'onofrio MVF, Howard BV (1996) Does free fatty acid infusin impair insulin action also through an increase in oxidative stress? J Clin Endocrinol Metab 81:4244–4248

Pirola L, Johnston AM, Obberghen EV (2004) Modulation of insulin action. Diabetologia 47:170–184

Rathz DA, Brown KM, Kramer LA, Liggett SB (2002) Amino acid 49 polymorphism in the human beta-1 adrenergic receptor affect agonist-promoted trafficking. J Cardiovasc Pharmacol. 30:155–160

Reynisdottir S, Wahrenberg H, Carlström K, Rössner S, Arner P (1994) Catecholamine resistance in fat cells of women with upper body obesity due to decreased expression of beta2-adrenoceptors. Diabetologia 37:428–435

Ribarik Coe N, Bernlohr DA (1998) Physiological properties and functions of intracellular fatty acid-binding proteins. Biochim Biophys Acta 1391:287–306

Richards M, A, Lainchbury JG, Nicholls MG, Troughton RW, Yandle TG (2002) BNP in hormone-guided treatment of heart failure. Trends Endocrinol Metab 13:151–155

Richelsen B (1986) Increased alpha2- but similar beta-adrenergic receptor activities in subcutaneous gluteal adipocytes from females compared with males. Eur J Clin Invest 16:302–309

Romijn JA, Fliers E (2005) Sympathetic and parasympathetic innervation of adipose tissue: metabolic implications. Curr Opin Clin Nutr Metab Care 8:440–444

Ryden M, Dicker A, van Harmelen V, Hauner H, Brunnberg M, Perbcck L, Lönnqvist F, Arner P (2002) Mapping of early signaling events in tumor necrosis-alpha-mediated lipolysis in human fat cells. J Biol Chem 277:1085–1091

Schiffelers SLH, Blaak EE, Saris WHM, Baak MAv (2000) In vivo β3-adrenergic stimulation of human thermogenesis and lipid utilization. Clin Pharmacol Therap 67:558–566

Schiffelers SLH, Saris WHM, Boomsma F, Baak MAv (2001) Beta1- and Beta2-adrenoceptor-mediated thermogenesis and lipid utilization in obese and lean men. J Clin Endocrinol Metab 86:2191–2199

Schutz Y, Flatt JP, Jéquier E (1989) Failure of dietary fat intake to promote fat oxidation: a factor favoring the development of obesity. Am J Clin Nutr 50:307–314

Sengenes C, Berlan M, de Glisezinski I, Lafontan M, Galitzky J (2000) Natriuretic peptides: a new lipolytic pathway in human adipocytes. FASEB J 14:1345–1351

Sengenes C, Zakaroff-Girard A, Moulin A, Berlan M, Bouloumié A, Lafontan M, Galitzky J (2002) Natriuretic peptide-dependent lipolysis in fat cells is a primate specificity. Am J Physiol 283:R257–R265

Sengenes C, Bouloumié A, Hauner H, Berlan M, Busse R, Lafontan M (2003) Involvement of a cGMP-dependent pathway in natriuretic peptide-mediated hormone-sensitive lipase phosphorylation in human adipocytes. J Biol Chem 278:48617–48626

Shen W-J, Sridhar K, Bernlohr DA, Kraemer FB (1999) Interaction of rat hormone-sensitive lipase with adipocyte lipid-binding protein. Proc Natl Acad Sci USA 96:5528–5532

Shihara N, Yasuda K, Moritani T, Ue H, Adachi T, Tanaka H, Tsuda K, Seino Y (1999) The association between Trp64Arg polymorphism of the beta3-adrenergic receptor and autonomic nervous system activity. J Clin Endocrinol Metab 84:1623–1627

Siegel AJ, Lewandrowsky EL, Chun KY, Sholar MB, Fishman AJ, Lewandrowsky KB (2001) Changes in cardiac markers including B-natriuretic peptide in runners after the Boston marathon. Am J Cardiol 88:920–923

Smith CJ, Manganiello VC (1989) Role of hormone-sensitive low Km cAMP phosphodiesterase in regulation of cAMP-dependent protein kinase and lipolysis in rat adipocytes. Mol Pharmacol 35:381–386

Soni KG, Lehner R, Metalnikov P, O'Donnell P, Semache M, Gao W, Ashman K, Pshezhetsky AV, Mitchell GA (2004) Carboxylesterase 3 (EC 3.1.1.1.) is a major adipocyte lipase. J Biol Chem 279:40683–40689

Stich V, de Glisezinsky I, Crampes F, Hejnova J, Cottet-Emard J-M, Galitzky J, Lafontan M, Rivière D, Berlan M (2000) Activation of alpha2-adrenergic receptors impairs exercise-induced lipolysis in SCAT of obese subjects. Am J Physiol 279:R499–R504

Stich V, Marion-Latard F, Hejnova J, Viguerie N, Lefort C, Suljkovicova H, Langin D, Lafontan M, Berlan M (2002) Hypocaloric diet reduces exercise-induced alpha2-adrenergic antilipolytic effect and alpha2-adrenergic receptor mRNA levels in adipose tissue of obese women. J Clin Endocrinol Metab 87:1274–1281

Strålfors P, Belfrage P (1985) Phosphorylation of hormone-sensitive lipase by cyclic GMP-dependent protein kinase. FEBS Lett 180:280–284

Sullivan JE, Brocklehurst KJ, Marley AE, Carey F, Carling D, Beri RK (1994) Inhibition of lipolysis and lipogenesis in isolated rat adipocytes with AICAR, a cell-permeable activator of AMP-activated protein kinase. FEBS Lett 353:33–36

Taggart AKP, Kero J, Gan X, Cai T-Q, Cheng K, Ippolito M, Ren N, Kaplan R, Wu K, Wu T-J, Jin L, Liaw C, Chen R, Richman J, Connolly D, Offermans S, Wright SD, Waters G (2005) (D)-β-hydroxybutyrate inhibits adipocyte lipolysis via the nicotininc acid receptor PUMA-G. J Biol Chem 280:26649–26652

Tansey JT, Sztalryd C, Hlavin EM, Kimmel AR, Londos C (2004) The central role of perilipin A in lipid metabolism and adipocyte lipolysis. IUBMB Life 56:379–385

Tavernier G, Barbe P, Galitzky J, Berlan M, Caput D, Lafontan M, Langin D (1996) Expression of β3-adrenoceptors with low lipolytic action in human subcutaneous fat cells. J Lipid Res 37:87–97

Tavernier G, Jimenez M, Giacobino JP, Hulo N, Lafontan M, Muzzin P, Langin D (2005) Norepinephrine induce lipolysis in beta1/beta2/beta3-adrenoceptor knock out mice. Mol Pharmacol 68:793–799

Tiraby C, Langin D (2003) Conversion from white to brown adipocytes: a strategy for the control of fat mass. Trends Endocrinol Metab 14:439–441

Tunaru S, Kero J, Schaub A, Wufka C, Blaukat A, Pfeffer K, Offermanns S (2003) PUMA-G and HM74 are receptors for nicotinic acid and mediate its anti-lipolytic effect. Nature Med 9:352–355

Van Pelt RE, Jankowski CM, Gozansky WS, Schwartz RS, Kohrt WM (2005) Lower body adiposity and metabolic protection in postmenopausal women. J Clin Endocrinol Metab. 90:4573–4578

Villena JA, Roy S, Sarkady-Nagy E, Kim K-H, Sul HS (2004) Desnutrin, an adipocyte gene encoding a novel patatin domain-containing protein, is induced by fasting and glucocorticoids: ectopic expression of desnutrin increases triglyceride hydrolysis. J Biol Chem 279:47066–47075

Wahrenberg H, Lönnqvist F, Arner P (1989) Mechanisms underlying regional differences in lipolysis in human adipose tissue. J Clin Invest 84:458–467

Wajchenberg BL (2000) Subcutaneous and visceral adipose tissue: their relation to the metabolic syndrome. Endocrine Rev 21:697–738

Wise A, Foord SM, Fraser NJ, Barnes AA, Elshourbagy N, Eilert M, Ignar DM, Murdock PR, Steplewski K, Green A, Brown AJ, Dowell SJ, Szekeres PG, Hassal DG, Marshall FH, Wilson S, Pike NB (2003) Molecular identification of high and low affinity receptors for nicotinic acid. J Biol Chem 278:9869–9874

Wu X, Hoffstedt J, Deeb W, Singh R, Sedkova N, Zilbering A, Zhu L, Park PK, Arner P, Goldstein B (2001) Depot-specific variation in protein-tyrosine phosphatase activities in human omental and subcutaneous adipose tissue: a potential contribution to differential insulin sensitivity. J Clin Endocrinol Metab 86:5973–5980

Yamada K, Ishiyama-Shigemoto S, Ichikawa F, Yuan X, Koyanagi A, Koyama W, Nonaka K (1999) Polymorphism in the 5′-leader cistron of the beta2-adrenergic receptor gene associated with obesity and type 2 diabetes. J Clin Endocrinol Metab 84:1754–1757

Zechner R, Strauss JG, Haemmerle G, Lass A, Zimmermann R (2005) Lipolysis: pathway under construction. Curr Opin Lipidol 16:333–340

Zhang HH, Halbleib M, Ahmad F, Manganiello VC, Greenberg AS (2002) Tumor necrosis factor-α stimulates lipolysis in differentiated human adipocytes through activation of extracellular signal-related kinase and elevation of intracellular cAMP. Diabetes 51:2929–2935

Zhang J, Hupfeld CJ, Taylor SS, Olefsky JM, Tsien RY (2005) Insulin disrupts β-adrenergic signalling to protein kinase A in adipocytes. Nature 437:569–573

Zierath J, Livingston J, Thorne A, Bolinder J, Reynisdottir S, Lonnqvist F, Arner P (1998) Regional difference in insulin inhibition of non-esterified fatty acid release from human adipocytes: relation to insulin receptor phosphorylation and intracellular signalling through the insulin receptor substrate-1 pathway. Diabetologia 41:1343–1354

Zimmermann R, Strauss JG, Haemmerle G, Schoiswohl G, Birner-Gruenberger R, Riederer M, Lass A, Neuberger G, Eisenhaber F, Hermetter A, Zechner R (2004) Fat mobilization in adipose tissue is promoted by adipose triglyceride lipase. Science 306:1383–1386

Zurlo F, Lillioja S, Puente AED, Nyomba BL, Raz I, Saad MF, Swinburn BA, Knowler WC, Bogardus C, Ravussin E (1990) Low ratio of fat to carbohydrate oxidation as predictor of weight gain: study of 24-h RQ. Am J Physiol 259:E650–E657

increases cardiac contractility and reduces cardiac loading. Apelin may therefore play a crucial role in the control of body fluid homeostasis and cardiovascular functions. A clinical study in healthy volunteers to determine whether apelin controls water balance in humans is in progress. If this hypothesis is confirmed, the development of non-peptide agonists of the apelin receptor may therefore represents new therapeutic avenues for the treatment of water/sodium retention, heart and kidney failure.

Introduction

During the last decade, we have focused our work on the study of the organization and the functional role of the brain renin-angiotensin system (RAS). We first discovered that aminopeptidase A (EC 3.4.11.7; APA) and aminopeptidase N (EC 3.4.11.2; APN), two membrane-bound zinc metalloproteases, are involved in vivo in the metabolism of brain angiotensin II (AngII) and angiotensin III (AngIII), respectively (Zini et al. 1996). Then, using specific and selective APA and APN inhibitors (Chauvel et al. 1994), we demonstrated that AngIII and not AngII, as established at the periphery, is the major effector peptide of the brain RAS, exerting a tonic stimulatory effect on arterial blood pressure (BP) in hypertensive animals. In addition to stimulating BP (Reaux et al. 1999), AngIII triggers vasopressin (AVP) release and activates magnocellular vasopressinergic neurons in the supraoptic nucleus (SON) when applied locally in this structure (Zini et al. 1996; 1998). The later effects, however, are not mediated by conventional type-1 (AT_1) or type-2 (AT_2) angiotensin receptors (AT), since neither mRNA expression nor binding sites corresponding to AT_1 or AT_2 have been found in the SON (Allen et al. 1992; Lenkei et al. 1997). Therefore, the nature of the receptor involved in the action of AngIII on AVP release remained to be defined.

Searching for a receptor for AngIII, we cloned, by homology from a rat brain cDNA library, an orphan GPCR, the rat APJ receptor. To identify the endogenous ligand of this orphan GPCR, we developed a new process based on the direct observation of ligand-induced internalization of fluorescently tagged receptors expressed in mammalian cells after incubation with prepurified tissue extracts. Subsequently, the endogenous ligand of the human APJ receptor was isolated by Tatemoto et al. (1998) and named apelin. We then studied the distribution and the physiological role of this new peptide.

Search for a putative angiotensin receptor subtype specific for AngIII in the CNS

In an attempt to isolate a putative angiotensin receptor subtype specific for AngIII, we screened SON RNA by reverse transcription and PCR amplification using primers located in the consensus transmembrane domains of the rat angiotensin receptor subtypes AT_{1A} and AT_{1B}. We obtained a PCR fragment of the sequence that showed an 35% amino acid identity with the rat AT_{1A} receptor and which was subsequently used to screen a rat brain cDNA library. We then isolated an open reading frame of 1 134 bp encoding a seven-transmembrane domain, GPCR composed of 377 amino acids (De Mota et al. 2000; Fig. 1).

Alignment of the coding sequence shows a high percentage of identical amino acids (between 90 and 96%) with those of the mouse (Devic et al. 1999) and human (O'Dowd

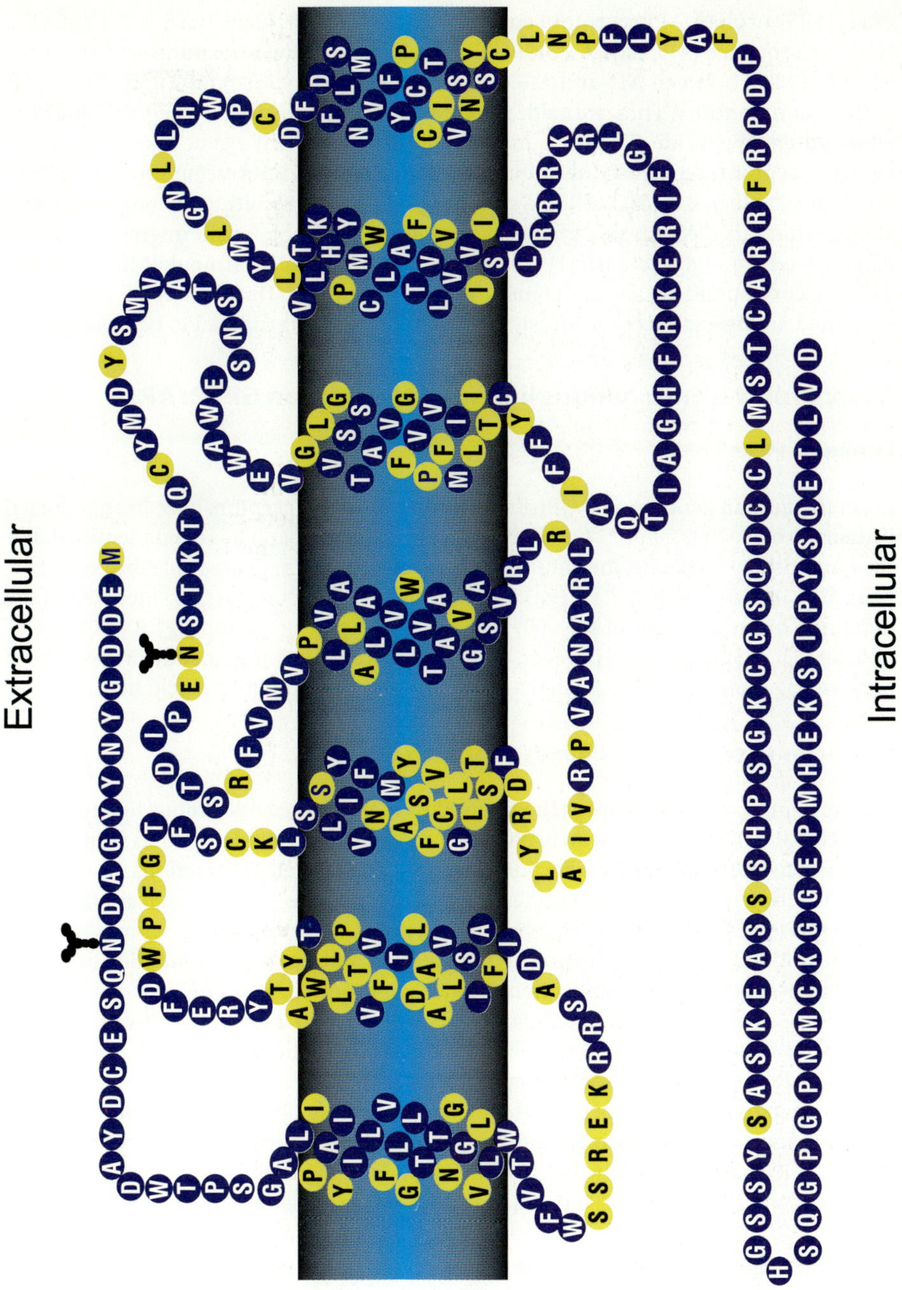

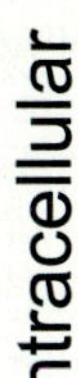

Fig. 1. Serpentine representation of the rat apelin receptor. Identical amino acids between apelin receptor and the rat AT$_{1A}$ receptor are shown in *yellow*

et al. 1993) orphan APJ receptor sequences and only 31% identity with the rat AT_{1A} receptor sequence (Murphy et al. 1991; Fig. 1). In spite of the low amino acid sequence identity (32%) between AT_1 and AT_2 receptors, these receptors bind AngII and AngIII with the same affinity. Thus, we made the hypothesis that our cloned receptor could bind an angiotensin peptide. To verify this hypothesis, we created a gene encoding the rat cloned receptor fused at its C-terminus to enhanced green fluorescent protein (EGFP). We then established a CHO cell line stably expressing the resulting fusion protein and showed that this receptor was unable to bind iodinated angiotensin fragments such as AngII, AngIII and angiotensin IV (AngIV), thereby demonstrating that this receptor did not correspond to an angiotensin receptor subtype (De Mota et al. 2000) and remained an orphan receptor for which the endogenous ligand had to be isolated.

Search for the endogenous ligand of the orphan GPCR: APJ

Orphan GPCRs

From the human genome, 367 GPCRs excluding sensory receptors have been reported (Vassilatis et al. 2003). For 224 of them, the endogenous ligand has been identified. For the remaining 143 GPCRs, the natural ligand is unknown; they constitute the pool of orphan GPCRs. Orphan GPCRs represent interesting therapeutic targets since 50% of the marketed drugs and 30% of the 500th top best-selling drugs act on GPCRs (Katugampola and Davenport 2003). However, these GPCRs represent therapeutic targets only if their endogenous ligand is isolated and their biological functions elucidated.

Strategies used to isolate ligands of orphan GPCRs

A reverse pharmacology approach was used to identify the endogenous ligands of orphan GPCRs. In this strategy, an orphan receptor is stably transfected into mammalian cells, which are then exposed to purified reverse phase HPLC fractions from tissue extracts, containing natural peptides that might mediate their biological effects via binding and activation of the transfected receptor. Activation of the orphan receptor by its ligand results in the activation of transduction pathways and production of second messenger responses as mobilization of calcium, production of cAMP or arachidonic acid (Wise et al. 2002). These cellular events result in the metabolic activation of the cells that lead to a release of protons in the extracellular milieu. This acidification can be monitored using a cytosensor microphysiometer (McConnell et al. 1992). Since 1995, only a group of fifteen new peptidic sequences, including nociceptin, leucokinin, orexins, apelin, prolactin-releasing peptide, ghrelin, motilin, metastatin, neuropeptides B and W, 26RFa, and Obestatin, have been isolated (Katugampola and Davenport 2003; Chartrel et al. 2003; Zhang 2005), emphasizing the difficulty of this research field.

Disadvantages of the strategies devised

Signal transduction-based approaches have the disadvantage of being highly dependent on the successful prediction of the transduction pathways used by the orphan GPCR. In addition, the orphan receptor may have to be expressed in a variety of cell lines before one is found to exhibit viable coupling (Stadel et al. 1997). To circumvent these shortcomings, the orphan GPCR is transfected into mammalian cells in the presence of one or more of a cocktail of promiscuous G-proteins, such as $G\alpha_{16/15}$ (Offermanns and Simon 1995), or with chimeric G-proteins, such as $G\alpha_{qi5}$ or $G\alpha_{qs5}$ (Conklin et al. 1996), in which the N-terminal five amino acids of $G\alpha_q$ have been replaced with the corresponding amino acids of the $G\alpha_i$ or $G\alpha_s$ to facilitate receptor coupling to calcium mobilization (Wise et al. 2002). However, the greatest difficulty remained: both signal transduction-based approaches and the more generic cytosensor microphysiometer assay are impeded by the presence of many endogenous GPCRs in the host cell line, which may interfere with the recorded signal.

A new process based on internalization

A novel screening process to identify ligands of orphan GPCRs was developed. This method is based on the property of GPCRs to internalize upon ligand exposure (Lenkei et al. 2000). The process involves the fluorescent tagging of the orphan GPCR, the expression of the tagged receptor at the surface of eucaryotic cells and the incubation of the transfected cells with fractions purified from tissue extracts. The advantages of this procedure are the independence of signaling mechanisms downstream of the receptor, the direct visualization of the studied target and the absence of background. This method is highly sensitive (50 fmoles in 10 µl), applicable to a majority of GPCRs.

We first validated our procedure by recovering the endogenous ligand of the neurotensin type 1 (NT_1) receptor from reversed-phase HPLC fractions of frog brain extract (Lenkei et al. 2000). We then used this procedure to isolate the endogenous ligand of the orphan rat APJ receptor. For this purpose, we used the previously established stable CHO cell line expressing the rat APJ receptor tagged at its C-terminal part with the EGFP to evaluate the ability of the 120 reversed-phase HPLC fractions obtained from purification of 2,500 frog brains to induce the internalization of the rat APJ-EGFP receptor. Only the fourth pool (composed of fractions 16 to 20) of the 120 frog brain fractions tested on the rat APJ-EGFP receptor-expressing CHO cells induced receptor internalization, as monitored by confocal microscopy (Fig. 2a).

Then the fractions of the fourth pool were evaluated individually and only the fractions 18 and 19 promoted the receptor internalization (Fig. 2b). In a parallel study, Tatemoto et al. isolated at the end of 1998 the endogenous ligand of the human APJ receptor by using the cytosensor microphysiometer method, which they named apelin for APJ endogenous ligand (Tatemoto et al. 1998).

Structure of the precursor mRNA/gene
and processing of the precursor

Apelin is a 36 amino acid peptide (apelin 36) generated from a 77 amino acids precursor, the proapelin (Fig. 3a). This precursor has been isolated from various species (Tatemoto et al. 1998; Habata et al. 1999; Lee et al. 2000). The human proapelin gene is located on chromosome X at locus Xq25-q26.1 and contains three exons, with the coding region spanning exons 1 and 2. The 3′ untranslated region also spans two exons (2 and 3). This fact may account for the presence of transcripts of two different sizes ($\sim$ 3 kb and $\sim$ 3.6 kb) in various tissues (O'Carroll et al. 2000; Lee et al. 2000). The alignment of

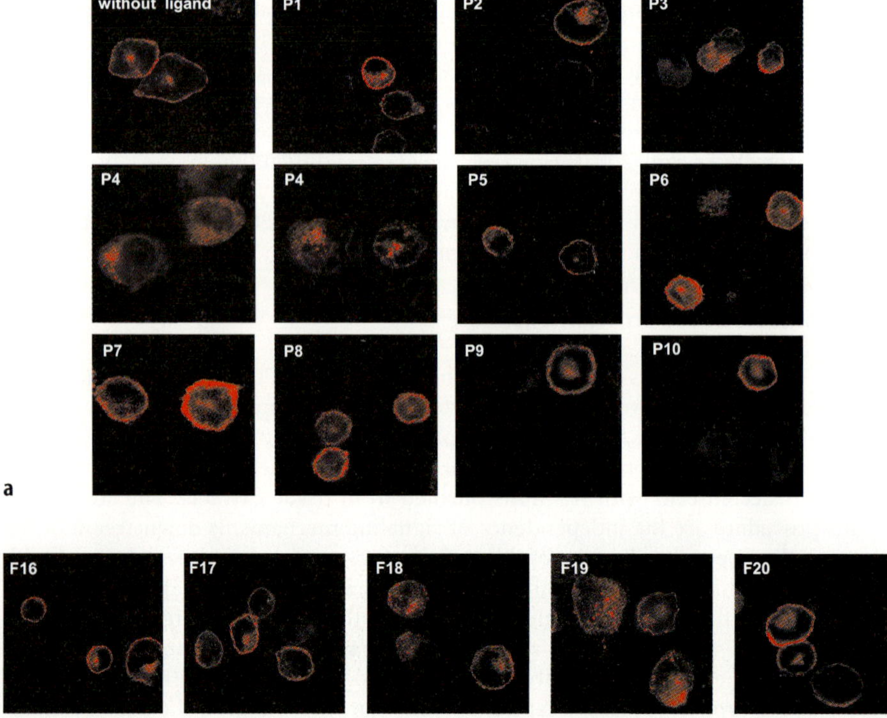

Fig. 2. Internalization of the rat APJ-EGFP receptor stably transfected in CHO cells monitored by confocal microscopy. (**a**) Internalization was tested with the first 10 pools of the 120 frog brain reversed-phase HPLC fractions at the first step of purification (each pool was composed of five consecutive fractions). (**b**) Internalization was tested with the five consecutive fractions, F16, F17, F18, F19 and F20, which compose pool 4

proapelin amino acid sequences from cattle, human, rat, and mouse has demonstrated strict conservation of the last C-terminal 17 amino acids, known as apelin-17 or K17F (Fig. 3a). In vivo, proapelin gives rise to various molecular forms of apelin, probably through the action of prohormone convertases due to the presence of pairs of basic residues in proapelin. In rat brain and plasma, the predominant forms of apelin are the pyroglutamyl form of apelin 13 (pE13F) and, to a lesser extent, K17F (Fig. 3b; De Mota et al. 2004). In rat lung, testis, and uterus and bovine colostrum apelin 36 predominates whereas, in the rat mammary gland, both apelin 36 and pE13F have been detected (Hosoya et al. 2000; Kawamata et al. 2001).

Distribution of apelin and its receptor in the rat brain

The production of a polyclonal antibody with high affinity and selectivity for K17F (De Mota et al. 2004) has made it possible to visualize, for the first time, apelin neurons in the rat central nervous system. The precise central topographical distribution

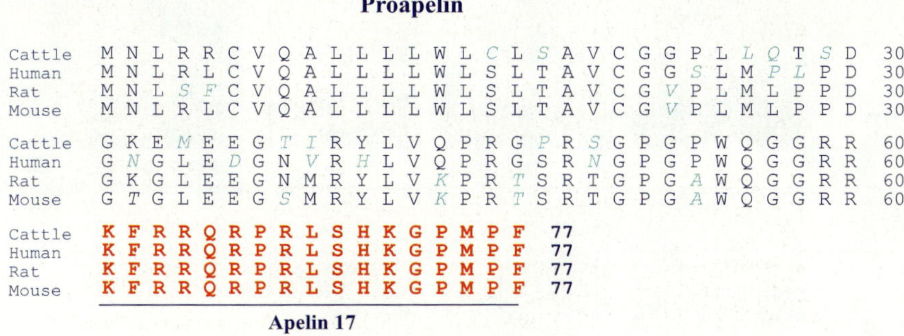

Fig. 3. (a) Amino acid sequence of the apelin precursor, proapelin, in cattle, humans, rats, and mice. The amino acid sequence of apelin 17 is underlined. (b) Apelin fragments detected in vivo in mammals: apelin 17 (K17F) and the pyroglutamyl form of apelin 13 (pE13F)

of apelin immunoreactivity shows that apelin-immunoreactive (IR) cell bodies are particularly abundant in the structures of the hypothalamus and medulla oblongata that are involved in neuroendocrine control, drinking behavior, and the regulation of arterial BP, notably in the hypothalamic SON and the magnocellular part of the paraventricular nucleus (PVN), the arcuate nucleus, the lateral reticular nucleus, and the nucleus ambiguus (Fig. 4a; Reaux et al. 2002). Conversely, apelin-IR nerve fibers are much more widely distributed in many brain regions than neuronal apelin cell bodies. The density of IR nerve fibers and apelinergic nerve endings is highest in the inner layer of the median eminence and in the posterior pituitary (Reaux et al. 2001; Brailoiu et al. 2002), suggesting that the apelin neurons of the SON and PVN, like the magnocellular AVP and ocytocin neurons, project into the posterior pituitary. Double immunofluorescence staining confirmed this finding, showing that apelin co-localized with AVP (Reaux-Le Goazigo et al. 2004; De Mota et al. 2004) and ocytocin (Brailoiu et al. 2002) in magnocellular hypothalamic neurons.

Apelin-IR nerve fibers also innervate the mesencephalon, the pons, the medulla oblongata, and several circumventricular organs such as the vascular organ of the lamina terminalis (OVLT), the subfornical organ (SFO), the subcommissural organ, and the area postrema (Reaux et al. 2002).

Like apelin, the apelin receptor is widely distributed throughout the rat central nervous system (O'Carroll et al. 2000; Lee et al. 2000; De Mota et al. 2000). In situ hybridization studies have shown that apelin receptor mRNA is present in the piriform and entorhinal cortices, the septum, the hippocampus and structures containing monoaminergic neuronal cell bodies (pars compacta of the substantia nigra, dorsal

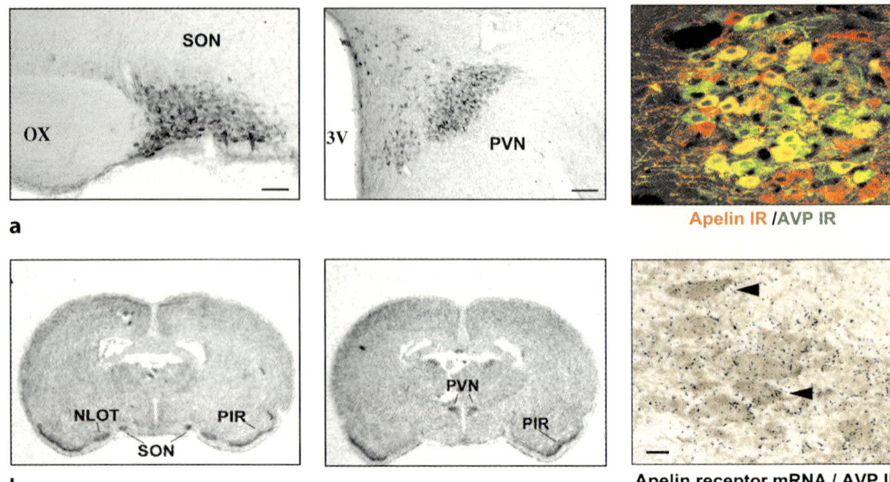

Fig. 4. Rat hypothalamic distribution of apelin immunoreactivity and apelin receptor mRNA. (a) Distribution of apelin-immunoreactive (IR) nerve cell bodies in the rat supraoptic nucleus (SON) and paraventricular nucleus (PVN) scale bar = 100 µm. ox, optic chiasma; 3v, third ventricle. Numerous apelin neurons are detected in both nuclei and, in the PVN, apelin (*red*) co-localizes with vasopressin (*green*). (**b**) Distribution of apelin receptor mRNA by in situ hybridization in the rat SON and PVN. Apelin receptor mRNA is highly expressed in both nuclei and synthesized in magnocellular vasopressin neurons (*arrowheads*) scale bar = 100 µm. Abbreviations: nucleus of the lateral olfactory tract (NLOT), piriform cortex (PIR). Adapted from De mota et al. 2000; Reaux et al. 2002 and Reaux-Le Goazigo et al. 2004

raphe nucleus, and locus coeruleus). The apelin receptor is particularly abundant in the apelin-rich hypothalamic nuclei, including the SON, PVN and the arcuate nucleus, and in the pineal gland and the anterior and intermediate lobes of the pituitary gland (Fig. 4b; De Mota et al. 2000). Furthermore, double labeling studies combined with immunocytochemistry and in situ hybridization have demonstrated that, in the SON and PVN, apelin receptors (Reaux et al. 2001; O'Carroll et al. 2003), like type 1A and 1B AVP receptors (V_{1A} and V_{1B}; Hurbin et al. 1998), are synthesized by magnocellular AVP neurons, suggesting an interaction between AVP and apelin (Fig. 4b).

Involvement of apelin in the regulation of water balance

The neurosecretory neurons release AVP, an antidiuretic vasoconstrictor peptide, into the fenestrated capillaries of the posterior pituitary in response to changes in plasma osmolality and volemia (Manning et al. 1977; Brownstein et al. 1980). The recent report of co-localization of AVP and apelin in the magnocellular neurons of the hypothalamus and the presence of receptors for AVP and apelin on these same neurons (Fig. 5) suggest a potential apelinergic response to these stimuli.

Regarding the involvement of apelin in the regulation of water balance, it is possible that, independently of the feedback control exerted by AVP on its own release, apelin regulates AVP release. This hypothesis has been tested in lactating rats exhibit-

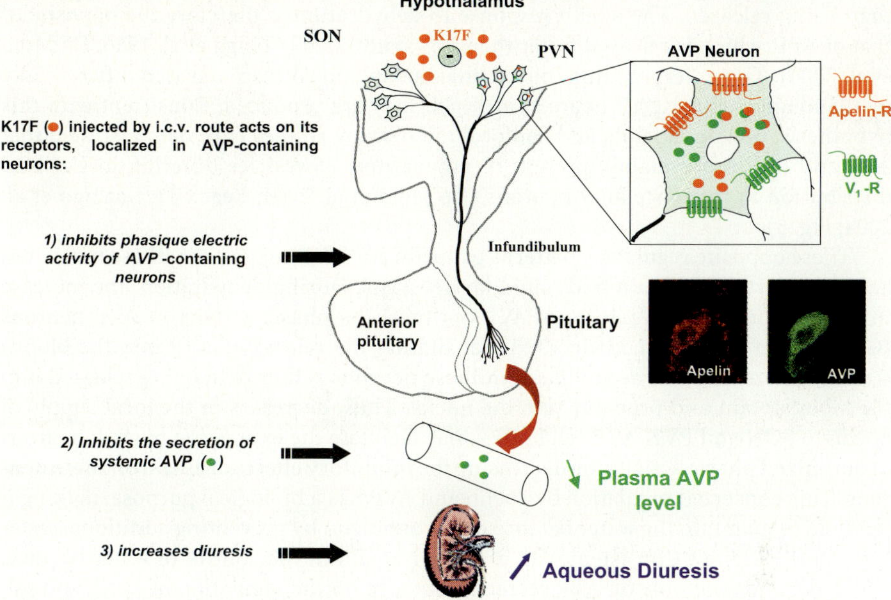

Fig. 5. Apelin, a potent diuretic neuropeptide that counteracts the effects of AVP through inhibition of AVP neuron activity and AVP release. In rodents, apelin and its receptor are co-localized with AVP in the SON and PVN magnocellular neurons. In lactating animals, the i.c.v. administration of apelin inhibited the phasic electrical activity of AVP neurons, thereby decreasing systemic AVP secretion and increasing water diuresis. Apelin-R: apelin receptor; V_1-R : vasopressin type 1 receptor. Adapted from Llorens Cortes et al. 2006

ing a reinforced phasic pattern of AVP neurons during lactation, thereby facilitating systemic AVP release to maintain body water content for optimal milk production. In this model, the intracerebroventricular (i.c.v.) injection of K17F inhibits the phasic firing activity of AVP neurons, thereby decreasing AVP release into the bloodstream, leading to aqueous diuresis (De Mota et al. 2004; Fig. 5). Similarly, a marked decrease in systemic AVP release is observed following the i.c.v. injection of K17F or pE13F in mice deprived of water for 24 h (Reaux et al. 2001), a condition known to increase AVP neuron activity. These data suggest that apelin is probably released from the SON and PVN AVP cell bodies and inhibits AVP neuron activity and release by acting directly on the apelin autoreceptors expressed by AVP/apelin-containing neurons. This mechanism probably involves apelin acting as a natural inhibitor of the antidiuretic effect of AVP. The co-localization and opposite biological actions of these two peptides raise questions concerning how these two peptides are regulated to maintain body fluid homeostasis. For this purpose, the effects of water deprivation on the neuronal content and release of both apelin and AVP were studied.

In rats deprived of water for 24 h, a large increase in the hypothalamic apelin content, especially in the PVN and SON magnocellular neurons (Reaux-Le Goazigo et al. 2004), is mirrored by a decrease in plasma apelin levels (De Mota et al. 2004), suggesting that, under these conditions, apelin accumulates within AVP neurons rather than being released. The apelin response to dehydration is therefore the opposite of that of AVP, which is released faster than it is synthesized (Zingg et al. 1986; De Mota et al. 2004). This interpretation implies that apelin and AVP are released differentially by the magnocellular AVP neurons in which they are produced. Consistent with this hypothesis, double-labeling and confocal microscopy studies have demonstrated that AVP and apelin are mainly present in populations of vesicles differing in size and distribution in magnocellular neurons (De Mota et al. 2004; Reaux-Le Goazigo et al. 2004; Fig. 5).

These opposite regulatory patterns of apelin and AVP suggest that these molecules act in concert to maintain body fluid homeostasis. During dehydration, the increase in the somatodendritic release of AVP optimize the phasic activity of AVP neurons (Gouzenes et al. 1998; Ludwig 1998), facilitating the release of AVP into the bloodstream, whereas apelin accumulates in these neurons rather than being released into the bloodstream and probably into the nuclei. Thus, decreases in the local supply of apelin to SON and PVN AVP cell bodies may facilitate the expression by AVP neurons of optimized phasic activity, by decreasing the inhibitory effects of apelin on these neurons. This concerted regulation by apelin and AVP has a biological purpose, making it possible to maintain the water balance of the organism by preventing additional water loss via the kidney. Consistent with a role for apelin in the control of water balance, which depends not only on AVP secretion but also on the regulation of water and salt intake, apelin administered i.c.v. clearly and significantly decreases water intake in rats deprived of water for 24 h (Reaux et al. 2001).

Peripheral cardiovascular actions

Apelin also has cardiovascular effects. The mRNA encoding apelin receptors has been detected in endothelial cells of large conduit arteries, coronary vessels, and the endocardium of the right atrium (O'Carroll et al. 2000; Kleinz and Davenport 2004). The injection of apelin into the bloodstream decreases arterial BP (Lee et al. 2000; Reaux et al. 2001; Tatemoto et al. 2001; El Messari et al. 2004), via a mechanism dependent on NO production (Tatemoto et al. 2001). In normotensive or hypertensive rats, apelin increases the contractile force of the myocardium via a positive inotropic effect (Chen et al. 2003; Berry et al. 2004). Moreover, APJ knockout mice display an enhanced vasopressor response to systemic AngII, suggesting a counter regulatory action of apelin on AngII (Ishida et al. 2004).

Acute administration of apelin in vivo results in a vasodilatation-induced decrease in left ventricular preload and afterload and a potent increase in contractility accompanied by a slight decrease in cardiac output. Conversely, chronic apelin infusion increases cardiac output without causing hypertrophy (Ashley et al. 2005). Apelin immunoreactivity has been found to increase in the plasma of patients in the early stages of heart failure and then to decrease during later, more severe stages of heart failure (Foldes et al. 2003).

These data suggest that apelin and its receptor could constitute potential therapeutic targets in the treatment of heart failure. Indeed, the administration of apelin or of a non-peptide agonist of the apelin receptor might improve the contractile performance of the myocardium while reducing cardiac loading and increasing aqueous diuresis in patients with heart failure.

Conclusions and pathophysiological implications

The identification of apelin as the endogenous ligand of the orphan APJ receptor constitutes a major advance, both for fundamental research and, potentially, for clinical practice. It demonstrates the validity of the "deorphanization" approach to orphan receptors for the identification of new bioactive peptides and new therapeutic targets. The experimental data obtained to date demonstrate that apelin, by inhibiting the phasic electrical activity of AVP neurons and the systemic secretion of AVP, induces aqueous diuresis. In the periphery, apelin decreases arterial BP and increases the contractile force of the myocardium. Overall, these data show that this newly identified circulating vasoactive peptide may play a key role in the maintenance of water balance and cardiovascular function. The development of non-peptide agonists of the apelin receptor, based on the knowledge of the structures of apelin and its receptor, could lead to new therapeutic tools for the treatment of the syndrome of inappropriate secretion of AVP, thirst disorders, and heart and kidney failure.

Acknowledgements. This work was funded by INSERM, the Société Française d'Hypertension Artérielle, the Fonds de la Recherche en Santé du Québec and the France-Hungary co-operation program BALATON.

References

Allen AM, Paxinos G, Song KF, Mendelsohn FAO (1992) Localization of angiotensin receptor binding sites in the rat brain. In: Björklund A, Hökfelt T, Kuhar MJ (eds) Handbook of chemical neuroanatomy: Neuropeptide receptors in the CNS. Elsevier, Amsterdam, p 1–37

Ashley EA, Powers J, Chen M, Kundu R, Finsterbach T, Caffarelli A, Deng A, Eichhorn J, Mahajan R, Agrawal R, Greve J, Robbins R, Patterson AJ, Bernstein D, Quertermous T (2005) The endogenous peptide apelin potently improves cardiac contractility and reduces cardiac loading in vivo. Cardiovasc Res 65:73–82

Berry MF, Pirolli TJ, Jayasankar V, Burdick J, Morine KJ, Gardner TJ, Woo YJ (2004) Apelin has in vivo inotropic effects on normal and failing hearts. Circulation 110:87–93

Brailoiu GC, Dun SL, Yang J, Ohsawa M, Chang JK, Dun NJ (2002) Apelin-immunoreactivity in the rat hypothalamus and pituitary. Neurosci Lett 327:193–197

Brownstein MJ, Russell JT, Gainer H (1980) Synthesis, transport, and release of posterior pituitary hormones. Science 207:373–378

Chartrel N, Dujardin C, Anouar Y, Leprince J, Decker A, Clerens S, Do-Rego JC, Vandesande F, Llorens-Cortes C, Costentin J, Beauvillain JC, Vaudry H (2003) Identification of 26RFa, a hypothalamic neuropeptide of the RFamide peptide family with orexigenic activity. Proc Natl Acad Sci USA 100:15247–15252

Chauvel EN, Coric P, Llorens-Cortes C, Wilk S, Roques BP, Fournie-Zaluski MC (1994) Investigation of the active site of aminopeptidase A using a series of new thiol-containing inhibitors. J Med Chem 37:1339–1346

Chen MM, Ashley EA, Deng DX, Tsalenko A, Deng A, Tabibiazar R, Ben-Dor A, Fenster B, Yang E, King JY, Fowler M, Robbins R, Johnson FL, Bruhn L, McDonagh T, Dargie H, Yakhini Z, Tsao PS, Quertermous T (2003) Novel role for the potent endogenous inotrope apelin in human cardiac dysfunction. Circulation 108:1432–1439

Conklin BR, Herzmark P, Ishida S, Voyno-Yasenetskaya TA, Sun Y, Farfel Z, Bourne HR (1996) Carboxyl-terminal mutations of Gq alpha and Gs alpha that alter the fidelity of receptor activation. Mol Pharmacol 50:885–890

De Mota N, Lenkei Z, Llorens-Cortes C (2000) Cloning, pharmacological characterization and brain distribution of the rat apelin receptor. Neuroendocrinology 72:400–407

De Mota N, Reaux-Le Goazigo A, El Messari S, Chartrel N, Roesch D, Dujardin C, Kordon C, Vaudry H, Moos F, Llorens-Cortes C (2004) Apelin, a potent diuretic neuropeptide counteracting vasopressin actions through inhibition of vasopressin neuron activity and vasopressin release. Proc Natl Acad Sci USA 101:10464–10469

Devic E, Rizzoti K, Bodin S, Knibiehler B, Audigier Y (1999) Amino acid sequence and embryonic expression of msr/apj, the mouse homolog of Xenopus X-msr and human APJ. Mech Dev 84:199–203

El Messari S, Iturrioz X, Fassot C, De Mota N, Roesch D, Llorens-Cortes C (2004) Functional dissociation of apelin receptor signaling and endocytosis: implications for the effects of apelin on arterial blood pressure. J Neurochem 90:1290–1301

Foldes G, Horkay F, Szokodi I, Vuolteenaho O, Ilves M, Lindstedt KA, Mayranpaa M, Sarman B, Seres L, Skoumal R, Lako-Futo Z, deChatel R, Ruskoaho H, Toth M (2003) Circulating and cardiac levels of apelin, the novel ligand of the orphan receptor APJ, in patients with heart failure. Biochem Biophys Res Commun 308:480–485

Gouzenes L, Desarmenien MG, Hussy N, Richard P, Moos FC (1998) Vasopressin regularizes the phasic firing pattern of rat hypothalamic magnocellular vasopressin neurons. J Neurosci 18:1879–1885

Habata Y, Fujii R, Hosoya M, Fukusumi S, Kawamata Y, Hinuma S, Kitada C, Nishizawa N, Murosaki S, Kurokawa T, Onda H, Tatemoto K, Fujino M (1999) Apelin, the natural ligand of the orphan receptor APJ, is abundantly secreted in the colostrum. Biochim Biophys Acta 13:25–35

Hosoya M, Kawamata Y, Fukusumi S, Fujii R, Habata Y, Hinuma S, Kitada C, Honda S, Kurokawa T, Onda H, Nishimura O, Fujino M (2000) Molecular and functional characteristics of APJ. Tissue distribution of mRNA and interaction with the endogenous ligand apelin. J Biol Chem 275:21061–21067

Hurbin A, Boissin-Agasse L, Orcel H, Rabie A, Joux N, Desarmenien MG, Richard P, Moos FC (1998) The V1a and V1b, but not V2, vasopressin receptor genes are expressed in the supraoptic nucleus of the rat hypothalamus, and the transcripts are essentially colocalized in the vasopressinergic magnocellular neurons. Endocrinology 139:4701–4707

Ishida J, Hashimoto T, Hashimoto Y, Nishiwaki S, Iguchi T, Harada S, Sugaya T, Matsuzaki H, Yamamoto R, Shiota N, Okunishi H, Kihara M, Umemura S, Sugiyama F, Yagami K, Kasuya Y, Mochizuki N, Fukamizu A (2004) Regulatory roles for APJ, a seven-transmembrane receptor related to angiotensin-type 1 receptor in blood pressure in vivo. J Biol Chem 279:26274–26279

Katugampola S, Davenport A (2003) Emerging roles for orphan G-protein-coupled receptors in the cardiovascular system. Trends Pharmacol Sci 24:30–35

Kawamata Y, Habata Y, Fukusumi S, Hosoya M, Fujii R, Hinuma S, Nishizawa N, Kitada C, Onda H, Nishimura O, Fujino M (2001) Molecular properties of apelin: tissue distribution and receptor binding. Biochim Biophys Acta 23:2–3

Kleinz MJ, Davenport AP (2004) Immunocytochemical localization of the endogenous vasoactive peptide apelin to human vascular and endocardial endothelial cells. Regul Pept 118:119–125

Lee DK, Cheng R, Nguyen T, Fan T, Kariyawasam AP, Liu Y, Osmond DH, George SR, O'Dowd BF (2000) Characterization of apelin, the ligand for the APJ receptor. J Neurochem 74:34–41

Lenkei Z, Palkovits M, Corvol P, Llorens-Cortes C (1997) Expression of angiotensin type-1 (AT1) and type-2 (AT2) receptor mRNAs in the adult rat brain: a functional neuroanatomical review. Front Neuroendocrinol 18:383–439

Lenkei Z, Beaudet A, Chartrel N, De Mota N, Irinopoulou T, Braun B, Vaudry H, Llorens-Cortes C (2000) A highly sensitive quantitative cytosensor technique for the identification of receptor ligands in tissue extracts. J Histochem Cytochem 48:1553–1564

Llorens Cortes C, Beaudet A (2005) Apelin, a new peptide that conteracts vasopressin secretion. Med Sci (Paris) 21:741–746

Ludwig M (1998) Dendritic release of vasopressin and oxytocin. J Neuroendocrinol 10:881–895

Manning M, Lowbridge J, Haldar J, Sawyer WH (1997) Design of neurohypophyseal peptides that exhibit selective agonistic and antagonistic properties. Fed Proc 36:1848–1852

McConnell HM, Owicki JC, Parce JW, Miller DL, Baxter GT, Wada HG, Pitchford S (1992) The cytosensor microphysiometer: biological applications of silicon technology. Science 257:1906–1912

Murphy TJ, Alexander RW, Griendling KK, Runge MS, Bernstein KE (1991) Isolation of a cDNA encoding the vascular type-1 angiotensin II receptor. Nature 351:233–236

O'Carroll AM, Selby TL, Palkovits M, Lolait SJ (2000) Distribution of mRNA encoding B78/apj, the rat homologue of the human APJ receptor, and its endogenous ligand apelin in brain and peripheral tissues. Biochim Biophys Acta 21:72–80

O'Carroll AM, Don AL, Lolait SJ (2003) APJ receptor mRNA expression in the rat hypothalamic paraventricular nucleus: regulation by stress and glucocorticoids. J Neuroendocrinol 15:1095–1101

O'Dowd BF, Heiber M, Chan A, Heng HH, Tsui LC, Kennedy JL, Shi X, Petronis A, George SR, Nguyen T (1993) A human gene that shows identity with the gene encoding the angiotensin receptor is located on chromosome 11. Gene 136:355–360

Offermanns S, Simon MI (1995) G alpha 15 and G alpha 16 couple a wide variety of receptors to phospholipase C. J Biol Chem 270:15175–15180

Reaux A, Fournie-Zaluski MC, David C, Zini S, Roques BP, Corvol P, Llorens-Cortes C (1999) Aminopeptidase A inhibitors as potential central antihypertensive agents. Proc Natl Acad Sci USA 96:13415–13420

Reaux A, De Mota N, Skultetyova I, Lenkei Z, El Messari S, Gallatz K, Corvol P, Palkovits M, Llorens-Cortes C (2001) Physiological role of a novel neuropeptide, apelin, and its receptor in the rat brain. J Neurochem 77:1085–1096

Reaux A, Gallatz K, Palkovits M, Llorens-Cortes C (2002) Distribution of apelin-synthesizing neurons in the adult rat brain. Neuroscience 113:653–662

Reaux-Le Goazigo AR, Morinville A, Burlet A, Llorens-Cortes C, Beaudet A (2004) Dehydration-induced cross-regulation of apelin and vasopressin immunoreactivity levels in magnocellular hypothalamic neurons. Endocrinology 145:4392–4400

Stadel JM, Wilson S, Bergsma DJ (1997) Orphan G protein-coupled receptors: a neglected opportunity for pioneer drug discovery. Trends Pharmacol Sci 18:430–437

Tatemoto K, Hosoya M, Habata Y, Fujii R, Kakegawa T, Zou MX, Kawamata Y, Fukusumi S, Hinuma S, Kitada C, Kurokawa T, Onda H, Fujino M (1998) Isolation and characterization of a novel endogenous peptide ligand for the human APJ receptor. Biochem Biophys Res Commun 251:471–476

Tatemoto K, Takayama K, Zou MX, Kumaki I, Zhang W, Kumano K, Fujimiya M (2001) The novel peptide apelin lowers blood pressure via a nitric oxide-dependent mechanism. Regul Pept 99:87–92

Vassilatis DK, Hohmann JG, Zeng H, Li F, Ranchalis JE, Mortrud MT, Brown A, Rodriguez SS, Weller JR, Wright AC, Bergmann JE, Gaitanaris GA (2003) The G protein-coupled receptor repertoires of human and mouse. Proc Natl Acad Sci USA 100:4903–4908

Wise A, Gearing K, Rees S (2002) Target validation of G-protein coupled receptors. Drug Discov Today 7:235–246

Zhang JV, Ren PG, Avsian-Kretchmer O, Luo CW, Rauch R, Klein C, Hsueh AJ (2005) Obestatin, a peptide encoded by the ghrelin gene, opposes ghrelin's effects on food intake. Science 310:996–999

Zingg HH, Lefebvre D, Almazan G (1986) Regulation of vasopressin gene expression in rat hypothalamic neurons. Response to osmotic stimulation. J Biol Chem 261:12956–12959

Zini S, Fournie-Zaluski MC, Chauvel E, Roques BP, Corvol P, Llorens-Cortes C (1996) Identification of metabolic pathways of brain angiotensin II and III using specific aminopeptidase inhibitors: predominant role of angiotensin III in the control of vasopressin release. Proc Natl Acad Sci USA 93:11968–11973

Zini S, Demassey Y, Fournie-Zaluski MC, Bischoff L, Corvol P, Llorens-Cortes C, Sanderson P (1998) Inhibition of vasopressinergic neurons by central injection of a specific aminopeptidase A inhibitor. Neuroreport 9:825–828

Targeting regulators of G protein signaling (RGS proteins) to enhance agonist specificity

Richard R. Neubig[1]

Summary

Members of the diverse regulator of G protein signaling (RGS protein) family enhance the GTPase activity of G protein alpha subunits and speed their deactivation. Thus they negatively regulate signal transduction mediated by Gi- and Gq-coupled receptors. RGS proteins exhibit both tonic and regulated inhibition of agonist responses, differentially controlling the sensitivity of tissues depending on their post-translational modifications and expression levels. Reducing the activity of RGS proteins genetically or by means of chemical inhibitors can enhance G protein coupled receptor (GPCR) responses. RGS inhibitors present the novel possibility of enhancing agonist selectivity in a manner that depends on the signaling pathway employed or the tissue in which the receptor resides. To fully exploit this capability, more information will be needed about the expression of RGS proteins in different tissues and under distinct pathophysiological circumstances. Also, advances in the development of cell-permeable high affinity and selective inhibitors of specific RGS proteins will be needed. Finally, animal models illustrating the physiological functions of RGS proteins will be essential to predicting the actions in humans.

G protein coupled receptors (GPCRs) play a major role in signal transduction and are the targets of many therapeutic drugs. Signaling by G proteins is initiated by the agonist-mediated exchange of GTP for GDP on the Gα subunit and signaling is terminated by the hydrolysis of GTP to GDP followed by reassociation of the Gα and βγ subunits (Ross and Wilkie 2000). A recently described protein family, the regulator of G protein signaling (RGS) proteins, enhances the deactivation of the activity of G proteins and has now been show to play a major role in the control of GPCR signaling in vivo. In this report, I describe the mechanism and role of RGS proteins in controlling GPCR signaling and the potential utility of RGS inhibitors to enhance GPCR agonist responses.

The best known function of RGS proteins is to inhibit G protein signaling by accelerating GTP hydrolysis, thus turning off G protein signals (Berman et al. 1996). They are a highly diverse protein family, have unique tissue distributions, and are strongly regulated by signal transduction events (Hollinger and Hepler 2002; Neubig and Siderovski 2002). Also, evidence is emerging that, besides causing G protein inhibition, they can maintain efficient G protein signaling, serve as effectors to control downstream signals, and act as scaffold proteins to gather receptors, G proteins, effectors, and other regulatory molecules together (Siderovski et al. 1999; Hepler 2003).

[1] Department of Pharmacology, University of Michigan, Ann Arbor, Michigan

Conn et al.
Insights into Receptor Function
and New Drug Development Targets
© Springer-Verlag Berlin Heidelberg 2006

The RGS protein family

There are 20 classical RGS proteins and 10 related proteins containing RGS homology or RH domains (Hollinger and Hepler 2002). The RGS proteins are defined by the presence of an RGS domain, a compact $\sim$120 amino acid structure containing at least two connected 4-helix bundles. In addition to the RGS domain or "RGS box." many RGS proteins contain variable N- and C-terminal extensions that contain numerous signal transduction regulatory domains or scaffolds. The classical RGS proteins fall into four families based on the homology of their RGS domains, and they are named for one of their best-studied members (RZ, R4, R7, and R12). The N- and C-terminal extensions are often conserved within an RGS family.

Most members of the RZ and R4 families are quite small, with only the RGS domain plus a small, N-terminal extension that may play a role in targeting the RGS to the membrane or to receptors. The RZ family members (RGS 17, 19, and 20 – also known as Z2, GAIP, and Z1) have a cysteine-rich (cysteine string) motif in the N terminus. With the exception of RGS3, which has a large ($\sim$900 aa) but poorly understood N terminal extension, the R4 family members (RGS 1, 2, 3, 4, 5, 8, 13, 16, 18, and 21) all have a short N-terminal amphipathic sequence that has been implicated in receptor interactions (Zeng et al. 1998; Hague et al. 2005) and effector interactions (Salim et al. 2003). The R7 family members (RGS 6, 7, 9, 11) all have two additional domains. On the N-terminal side of the RGS domain is a dishevelled, EGL-10, pleckstrin (DEP) domain of unknown function, and C-terminal to the RGS domain is a G-protein gamma subunit-like (GGL) domain that forms a heterodimer with the unique G protein beta subunit, Gβ5 (Snow et al. 1999). While a mechanistic understanding of this complex remains elusive, it is essential to the stability and/or function of the R7 family proteins. A knock-out of either RGS9 or Gβ5 leads to the dramatic reduction of expression of both proteins in the retina, suggesting that they can only exist as a complex (Chen et al. 2003). The R12 family members differ most from each other, with RGS12 and 14 having two additional signal G protein interacting domains – a Rap binding domain that binds the small G protein rap with unknown consequences and a GoLoco or LGN domain that binds heterotrimeric Gα subunits and inhibits guanine nucleotide dissociation (i.e., a GDI function) (Hollinger and Hepler 2002). This latter domain may play a fundamental role in asymmetric cell division in early embryos as loss of RGS leads to embryonic lethality at the two-cell embryo stage (Martin-McCaffrey et al. 2004).

The interaction of the Gα subunit with these classical RGS proteins has been illuminated by the crystal structure of the complex of RGS4 with Gα_{i1} (Tesmer et al. 1997). Tight (nM affinity) binding only occurs in the presence of the GTPase transition state mimic GDP-AlF$_4^-$. The tight binding of RGS to the transition state for GTP hydrolysis provides a mechanistic explanation for the acceleration of GTP hydrolysis (i.e., lowered energy of the catalytic transition state). In this structure, the Gα subunit binds to the loops connecting helices 3 and 4, 5 and 6, and 7 and 8 of RGS4. This region of the RGS has been termed the "A" site for the site of Gα binding (Zhong and Neubig 2001); however, the RH domain-containing proteins (e.g., RhoGEF and GRK) utilize rather different surfaces for binding Gα subunits (Sterne-Marr et al. 2003; Chen et al. 2005).

The RH domain-containing proteins form a diverse collection and only some have been shown to interact with heterotrimeric G proteins. The RhoGEF proteins (p115rhoGEF, PDZ-rhoGEF, and leukemia-associated rhoGEF – LARG) bind to G12, G13, and, in some cases, Gq, and serve as effector molecules carrying the signal from the activated Gα subunit to the activation of the small G protein rho (Kozasa et al. 1998; Wang et al. 2004; Bian et al. 2005). This well-defined function is critical to the changes in cell shape, proliferation, and gene transcription mediated by G12/13 coupled receptors such as those for thrombin, lysophosphatidic acid (LPA), bombesin, etc. Interestingly, p115rhoGEF contains a structurally critical N-terminal extension of the RH domain. and it binds Gα$_{13}$ through contacts on RH domain loops α3–α4, α8–α9 and α10–α11 plus helix 8 as well as residues on the N-terminal extension (Chen et al. 2005). The rhoGEF RH domains can also GAP the Gα subunits but their primary role appears to be to serve as effectors, not as inhibitors of G12/13 signaling. The RH domains in proteins of the GPCR kinase or GRK family are specific for the Gαq family Gα subunits. Interestingly, this binding does not lead to increased GTP hydrolysis but it does strongly inhibit signaling by Gα$_q$ proteins via inhibition of effector binding. This intriguing protein family is able to inhibit all three signaling components of the GPCR system – the receptor, the Gα subunit, and the Gβγ subunit. The receptor is inhibited by phosphorylation by the GRK kinase domain and the Gα subunit (for Gq signaling only) is blocked by the RH domain, whereas the Gβγ subunit binds to the pleckstrin homology (PH) domain of GRK and is prevented from activating effectors (Lodowski et al. 2003). A recent report indicates that Axin binds Gα$_s$ to mediate activation of GSK3 and beta-catenin (Castellone et al. 2005). The other RH domain-containing proteins (AKAP2, SNX) have not been definitively shown to bind Gα subunits, so it is possible that the RH domain may play more of a structural or scaffolding role than one related to Gα signaling.

Physiologic Functions of RGS Proteins

A wide variety of in vitro studies both with purified RGS proteins and co-transfections into cell lines have shown that RGS proteins can strongly inhibit G protein signaling (see Hollinger and Hepler 2002; Traynor and Neubig 2005 for review). These studies have also illustrated the specificity of RGS proteins for different Gα subunits (Table 1). In addition to the specificity for Gα subunits, there is some evidence that RGS proteins may bind (directly or indirectly) to receptors and thus have more pronounced actions at those receptors, adding another level of specificity (Zeng et al. 1998; Wang et al. 2002; Hague et al. 2005).

Much less information is known about the physiological functions of endogenous RGS proteins in vivo. There are a number of knock-down studies using antisense RNA in vivo (Garzon et al. 2003, 2005; Sanchez-Blazquez et al. 2005), and highly informative mouse knock-out models have been reported for at least four RGS proteins. The RGS2 and RGS9 knock-outs have been studied most extensively. Loss of RGS2 was initially reported to have subtle immunologic and neurobehavioral effects (Oliveira-Dos-Santos et al. 2000) but was subsequently shown to have a dramatic hypertensive phenotype (Heximer et al. 2003) and loss of responsiveness to nitric oxide-mediated vasodilation (Tang et al. 2003). RGS9 knock-out mice have dramatically enhanced responses to drugs of abuse (Rahman et al. 2003; Zachariou et al. 2003). Kehrl and co-workers reported elegant studies on lymphocyte trafficking in RGS1 knock-out mice and show substantially altered B lymphocyte function and lymph node morphology (Moratz et al. 2004; Han et al. 2005). A recent report of an RGS4 knock-out shows subtle neurobehavioral effects (Grillet et al. 2005).

One potential difficulty with knock-outs of individual RGS proteins is evident from the dramatic redundancy of function shown in Table 1. Thus, the minimal phenotype of the RGS4 knock-out does not rule out a role for RGS4 but may simply be due to the presence of multiple RGS proteins that can act on $G\alpha_i$ and $G\alpha_q$ signaling. To assess the role of all RGS proteins in signaling by particular G proteins, we made use of a point mutation in $G\alpha$ subunits ($G\alpha_{i/o}$ G184S) that prevents the binding of and functional inhibition by RGS proteins (Lan et al. 1998; Fu et al. 2004). When expressed in cells, this mutation leads to dramatic enhancements of α_{2a} adrenergic (Jeong and Ikeda 2000), adenosine (Chen and Lambert 2000) and opioid signaling (Clark et al. 2003). To determine the full role of RGS proteins in the function of $G\alpha$ subunits, we recently prepared a genomic knock-in mouse model of the $G\alpha_{i2}^{G184S}$ allele. These mice show a dramatic pleiotropic phenotype with alterations in the cardiovascular, hematologic, and central nervous systems as well as alterations in bone and metabolic function (Huang et al., manuscript in preparation). This finding clearly shows that single RGS knock-outs do not reveal the full range of in vivo RGS functions. Specifically, these mice show a markedly increased bradycardic response to carbachol whereas the bradycardic effect of adenosine A1 receptors is virtually unchanged (Fu et al., submitted for publication). This finding also illustrates a point expanded upon below: inhibiting the action of RGS proteins, either genetically or pharmacologically, can produce a remarkable, selective potentiation of certain agonist responses while not influencing others.

Table 1. RGS specificity for different $G\alpha$ subunits. The ability of different RGS proteins to bind to or inhibit signaling by the indicated $G\alpha$ subunits – either in in vitro pull-down or GAP assays or in co-transfection studies – is indicated (for review see Hollinger and Hepler 2002; Traynor and Neubig 2005). RGS proteins highlighted in bold are selective for that particular G protein and have minimal activity at other G proteins

$G\alpha$ family	$G\alpha$ subunit	RGS family	RGS
$G\alpha_i$	$G\alpha_i$	R4, R12, RZ	1, 3, 4, 5, 8, 10, 12, 13, 14, 16, 17, 18, 19, 21
	$G\alpha_o$	R7	6, 7, 9, 11
	$G\alpha_t$	R7	9
	$G\alpha_z$	RZ	17, 19, 20
$G\alpha_s$	$G\alpha_s$	*	
$G\alpha_q$	$G\alpha_q$	R4, R12, RZ, GRK, RhoGEF	1, 2, 3, 4, 5, 8, 10, 12, 13, 14, 16, 17, 18, 19, 21 _GRK2_ LARG
$G_{\alpha12}$	$G\alpha_{12}$ & $G\alpha_{13}$	RhoGEF	_p115-RG, PDZ-RG, LARG_

* The RZ family members were previously known by other names: RGS17, RGS-Z2; RGS19, GAIP; and RGS20, RGS-Z1.

** There are no confirmed reports of an RGS protein binding $G\alpha_s$. SNX proteins were reported to GAP $G\alpha_s$ in 2001 (Zheng et al. 2001) but this has not been confirmed. A very recent report – also not confirmed – showed that the axin RH domain binds $G\alpha_s$ and mediates GSK3 and beta-catenin signaling (Castellone et al. 2005)

Potential of RGS proteins as Therapeutic Targets

Given the clear physiological role of RGS proteins, one could ask what the potential is for targeting RGS proteins therapeutically? In regard to the classical RGS proteins, preventing their action would likely increase signaling by endogenous hormones, neurotransmitters, or mediators. This depends on the prolongation of the active state of the G protein by RGS inhibition (Fig. 1). The specificity of these actions would depend on having inhibitors specific for individual RGS proteins but also on the tissue distribution of RGS proteins themselves. While much is known about RGS protein distribution in the brain (Gold et al. 1997; Grafstein-Dunn et al. 2001; Larminie et al. 2004), less is known about RGS expression and its regulation in peripheral endocrine tissues. This gap in knowledge must clearly be filled to better evaluate the potential of RGS inhibitors in endocrinology.

In addition to their use alone, RGS inhibitors could be combined with existing GPCR agonists to potentiate agonist action. Indeed, many GPCR agonists have a broad array of undesirable side effects accompanying their therapeutic action. In this context, RGS inhibitors would represent a novel class of therapeutics – "specificity enhancers". There are at least two ways in which RGS inhibitors could enhance the specificity of agonist drugs.

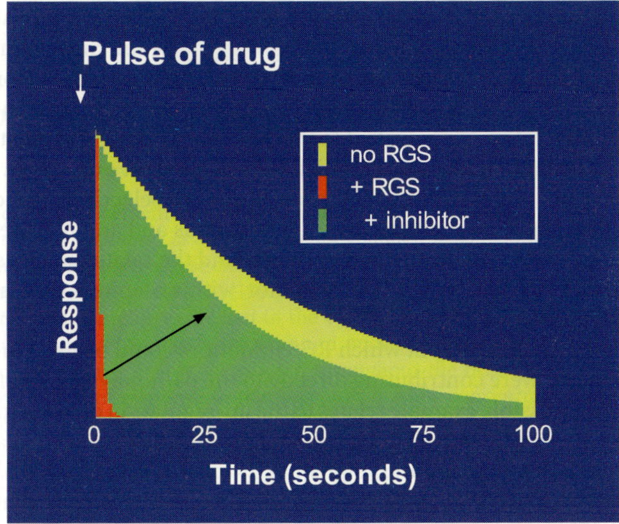

Fig. 1. RGS proteins speed the deactivation of Gα signals. RGS proteins serve as GTPase-activating proteins, or GAPs, and accelerate the deactivation of G protein signals. If there is no RGS protein present in a tissue and a short pulse of an agonist drug is applied, the G protein will be activated and then will turn off at a rate equal to the rate of intrinsic GTP hydrolysis by that G protein (*yellow curve* – typically a $t_{1/2}$ of about 20–30 seconds). If an RGS protein is present that speeds the GTP hydrolysis then the deactivation will be much faster (*red curve* $t_{1/2}$ of about 1 second or less). If the RGS protein that mediates this rapid deactivation were inhibited, it would substantially prolong the agonist signal (*green curve*) and would also lead to a markedly increased sensitivity of the tissue to that agonist

the activity of RGS proteins. They are regulated in numerous ways: transcriptional and post-transcriptional regulation of expression, post-translational modifications such as phosphorylation, membrane targeting, etc. One particular process that seems amenable to modulation is the control of RGS4 (and related RGS proteins) by acidic lipids and calmodulin. Wilkie and co-workers showed that phosphatidic acid and phosphatidyl inositol tris-phosphate can bind to RGS proteins and allosterically inhibit their GAP activity (Popov et al. 2000). Interestingly, this binding is reversed by the binding of Ca^{++}/calmodulin to RGS4. Ishii et al. (2002, 2005) have reported that this regulatory process may underlie the voltage and Ca^{++} dependent "relaxation" of GIRK currents. One could easily imagine a small-molecule compound that binds to the site on a RGS at which acidic phospholipids interact, thus reversing the inhibition and stimulating the RGS. There are a number of situations in which it might be desirable to enhance RGS action, the most obvious of which is suggested by the dramatic hypertension seen in the RGS2 knock-out mice. Enhancing RGS2 action could suppress signaling by many vasoconstrictor agonists and enhance the vasodilation mediated by nitric oxide and cGMP-dependent protein kinase.

Challenges in development of RGS-targeted therapeutics

Many challenges remain when considering RGS-targeted therapeutics. First, the physiology of RGS proteins outside of the brain and heart remains very poorly defined. Until the distribution of RGS protein expression in different tissues is characterized, and more importantly, the functional role of individual RGS proteins in a particular signaling processes is understood, it will be difficult to know which RGS proteins would make useful drug targets.

Second, given that RGS proteins act on their G protein target by a protein-protein interaction, the development of RGS inhibitors is likely to be more challenging than the development of well-defined enzymes or receptor targets. However, there has been a significant increase in the number of reports of successful protein-protein interaction inhibitor drugs in recent years (Berg 2003; Arkin and Wells 2004; Pagliaro et al. 2004 for review). Furthermore, our recent success in developing two single-digit micro-molar potency inhibitors – YJ34, an octapeptide (Jin et al. 2004) and, CCG-4986, a small-molecule compound (Roman and Neubig 2005) – also suggests that it should be feasible to overcome the problem of blocking the RGS/Ga protein-protein interaction.

Finally, a further difficulty may arise in the redundancy of RGS action at different G proteins. Table 1 shows that, for Gi and Gq in particular, a very large number of RGS proteins may inhibit their signaling, and inhibition of just one of them may have a very limited ability to potentiate signaling. This could lead to the need to develop "selectively non-selective" RGS inhibitors that can act on more than one RGS protein. Such a concept has been embraced for both GPCR drugs (e.g., antipsychotic agents) and tyrosine kinase inhibitors (Daub et al. 2004; Roth et al. 2004).

Future prospects

Despite the numerous challenges outlined above, the potential for RGS-targeted therapeutics to provide novel pharmacological actions (e.g., pathway- and tissue-related specificity enhancement) is exciting. It is clear that the priority for taking advantage of this potential must be a full understanding of the role of RGS proteins in the physiology of endocrine signaling. Second, novel chemistry and screening methods may be required to find potent RGS active compounds. The key role of GPCR-active drugs in the current armamentarium and the large number of agonists already available indicate that RGS protein modulation could find a significant place as a therapeutic approach.

Acknowledgements. Supported by NIH R01-GM39561. The author thanks Ying Fu, Xinyan Huang, David Roman, and Becky Roof for helpful discussions and for performing the studies that form the basis for this article.

References

Arkin MR, Wells JA (2004) Small-molecule inhibitors of protein-protein interactions: progressing towards the dream. Nat Rev Drug Discov 3:301–317

Berg T (2003) Modulation of protein-protein interactions with small organic molecules. Angew Chem Int Ed Engl 42:2462–2481

Berman DM, Wilkie TM, Gilman AG (1996) GAIP and RGS4 are GTPase-activating proteins for the Gi subfamily of G protein alpha subunits. Cell 86:445–452

Bian D, Mahanivong C, Yu J, Frisch SM, Pan ZK, Ye RD, Huang S (2005) The G(12/13)-RhoA signaling pathway contributes to efficient lysophosphatidic acid-stimulated cell migration. Oncogene

Castellone MD, Teramoto H, Williams BO, Druey KM, Gutkind JS (2005) Prostaglandin E2 promotes colon cancer cell growth through a Gs-axin-beta-catenin signaling axis. Science 310:1504–1510

Chen CK, Eversole-Cire P, Zhang H, Mancino V, Chen YJ, He W, Wensel TG, Simon MI (2003) Instability of GGL domain-containing RGS proteins in mice lacking the G protein beta-subunit Gbeta5. Proc Natl Acad Sci USA 100:6604–6609

Chen H, Lambert NA (2000) Endogenous regulators of G protein signaling proteins regulate presynaptic inhibition at rat hippocampal synapses. Proc Natl Acad Sci USA 97:12810–12815

Chen Z, Singer WD, Sternweis PC, Sprang SR (2005) Structure of the p115RhoGEF rgRGS domain-Galpha13/i1 chimera complex suggests convergent evolution of a GTPase activator. Nature Struct Mol Biol 12:191–197

Clark MJ, Harrison C, Zhong H, Neubig RR, Traynor JR (2003) Endogenous RGS protein action modulates mu-opioid signaling through Galphao. Effects on adenylyl cyclase, extracellular signal-regulated kinases, and intracellular calcium pathways. J Biol Chem 278:9418–9425

Daub H, Specht K, Ullrich A (2004) Strategies to overcome resistance to targeted protein kinase inhibitors. Nature Rev Drug Discov 3:1001–1010

Fu Y, Zhong H, Nanamori M, Mortensen RM, Huang X, Lan K, Neubig RR (2004) RGS-insensitive G-protein mutations to study the role of endogenous RGS proteins. Meth Enzymol 389:229–243

Garnier M, Zaratin PF, Ficalora G, Valente M, Fontanella L, Rhee MH, Blumer KJ, Scheideler MA (2003) Up-regulation of regulator of G protein signaling 4 expression in a model of neuropathic pain and insensitivity to morphine. J Pharmacol Exp Ther 304:1299–1306

Garzon J, Lopez-Fando A, Sanchez-Blazquez P (2003) The R7 subfamily of RGS proteins assists tachyphylaxis and acute tolerance at mu-opioid receptors. Neuropsychopharmacology 28:1983–1990

Garzon J, Rodriguez-Munoz M, Lopez-Fando A, Sanchez-Blazquez P (2005) The RGSZ2 protein exists in a complex with mu-opioid receptors and regulates the desensitizing capacity of Gz proteins. Neuropsychopharmacology 30:1632–1648

Gold SJ, Ni YG, Dohlman HG, Nestler EJ (1997) Regulators of G-protein signaling (RGS) proteins: region-specific expression of nine subtypes in rat brain. J Neurosci 17:8024–8037

Grafstein-Dunn E, Young KH, Cockett MI, Khawaja XZ (2001) Regional distribution of regulators of G-protein signaling (RGS) 1, 2, 13, 14, 16, and GAIP messenger ribonucleic acids by in situ hybridization in rat brain. Brain Res Mol Brain Res 88:113–123

Grillet N, Pattyn A, Contet C, Kieffer BL, Goridis C, Brunet JF (2005) Generation and characterization of Rgs4 mutant mice. Mol Cell Biol 25:4221–4228

Hague C, Bernstein LS, Ramineni S, Chen Z, Minneman KP, Hepler JR (2005) Selective inhibition of alpha1A-adrenergic receptor signaling by RGS2 association with the receptor third intracellular loop. J Biol Chem 280:27289–27295

Han SB, Moratz C, Huang NN, Kelsall B, Cho H, Shi CS, Schwartz O, Kehrl JH (2005) Rgs1 and Gnai2 regulate the entrance of B lymphocytes into lymph nodes and B cell motility within lymph node follicles. Immunity 22:343–354

Hepler JR (2003) RGS protein and G protein interactions: a little help from their friends. Mol Pharmacol 64:547–549

Heximer SP, Knutsen RH, Sun X, Kaltenbronn KM, Rhee MH, Peng N, Oliveira-dos-Santos A, Penninger JM, Muslin AJ, Steinberg TH, Wyss JM, Mecham RP, Blumer KJ (2003) Hypertension and prolonged vasoconstrictor signaling in RGS2-deficient mice. J Clin Invest 111:1259

Hollinger S Hepler JR (2002) Cellular regulation of RGS proteins: modulators and integrators of G protein signaling. Pharmacol Rev 54:527–559

Ishii M, Fujita S, Yamada M, Hosaka Y, Kurachi Y (2005) Phosphatidylinositol 3,4,5-trisphosphate and Ca2+/calmodulin competitively bind to the regulators of G-protein-signalling (RGS) domain of RGS4 and reciprocally regulate its action. Biochem J 385:65–73

Ishii M, Inanobe A, Kurachi Y (2002) PIP3 inhibition of RGS protein and its reversal by Ca2+/calmodulin mediate voltage-dependent control of the G protein cycle in a cardiac K+ channelProc Natl Acad Sci USA 99:4325–4330

Jeong SW, Ikeda SR (2000) Endogenous regulator of G-protein signaling proteins modify N-type calcium channel modulation in rat sympathetic neurons. J Neurosci 20:4489–4496

Jin Y, Zhong H, Omnaas JR, Neubig RR, Mosberg HI (2004) Structure-based design, synthesis, and pharmacologic evaluation of peptide RGS4 inhibitors. J Pept Res 63:141–146

Kozasa T, Jiang X, Hart MJ, Sternweis PM, Singer WD, Gilman AG, Bollag G, Sternweis PC (1998) p115 RhoGEF, a GTPase activating protein for Galpha12 and Galpha13. Science 280:2109–2111

Lan KL, Sarvazyan NA, Taussig R, Mackenzie RG, DiBello PR, Dohlman HG, Neubig RR (1998) A point mutation in Galphao and Galphai1 blocks interaction with regulator of G protein signaling proteins. J Biol Chem 273:12794–12797

Larminie C, Murdock P, Walhin JP, Duckworth M, Blumer KJ, Scheideler MA, Garnier M (2004) Selective expression of regulators of G-protein signaling (RGS) in the human central nervous system. Brain Res Mol Brain Res 122:24–34

Lodowski DT, Pitcher JA, Capel WD, Lefkowitz RJ, Tesmer JJ (2003) Keeping G proteins at bay: a complex between G protein-coupled receptor kinase 2 and Gbetagamma. Science 300:1256–1262

Martin-McCaffrey L, Willard FS, Oliveira-dos-Santos AJ, Natale DR, Snow BE, Kimple RJ, Pajak A, Watson AJ, Dagnino L, Penninger JM, Siderovski DP, D'Souza SJ (2004) RGS14 is a mitotic spindle protein essential from the first division of the mammalian zygote. Dev Cell 7:763–769

Moratz C, Hayman JR, Gu H, Kehrl JH (2004) Abnormal B-cell responses to chemokines, disturbed plasma cell localization, and distorted immune tissue architecture in Rgs1–/– mice. Mol Cell Biol 24:5767–5775

Neubig RR, Siderovski DP (2002) Regulators of G-protein signalling as new central nervous system drug targets. Nature Rev Drug Discov 1:187–197

Oliveira-Dos-Santos AJ, Matsumoto G, Snow BE, Bai D, Houston FP, Whishaw IQ, Mariathasan S, Sasaki T, Wakeham A, Ohashi PS, Roder JC, Barnes CA, Siderovski DP, Penninger JM (2000) Regulation of T cell activation, anxiety, and male aggression by RGS2. Proc Natl Acad Sci USA 97:12272–12277

Pagliaro L, Felding J, Audouze K, Nielsen SJ, Terry RB, Krog-Jensen C, Butcher S (2004) Emerging classes of protein-protein interaction inhibitors and new tools for their development. Curr Opin Chem Biol 8:442–449

Popov SG, Krishna UM, Falck JR, Wilkie TM (2000) Ca2+/Calmodulin reverses phosphatidylinositol 3,4, 5-trisphosphate-dependent inhibition of regulators of G protein-signaling GTPase-activating protein activity. J Biol Chem 275:18962–18968

Rahman Z, Schwarz J, Gold SJ, Zachariou V, Wein MN, Choi KH, Kovoor A, Chen CK, DiLeone RJ, Schwarz SC, Selley DE, Sim-Selley LJ, Barrot M, Luedtke RR, Self D, Neve RL, Lester HA, Simon MI, Nestler EJ (2003) RGS9 modulates dopamine signaling in the basal ganglia. Neuron 38:941–952

Roman DL, Neubig RR (2005) A novel multiplex flow cytometric method for measurement of Ga-RGS interaction and identification of small molecule inhibitors of RGS. FASEB J 19:259a

Ross EM, Wilkie TM (2000) GTPase-activating proteins for heterotrimeric G proteins: regulators of G protein signaling (RGS) and RGS-like proteins. Annu Rev Biochem 69:795–827

Roth BL, Sheffler DJ, Kroeze WK (2004) Magic shotguns versus magic bullets: selectively non-selective drugs for mood disorders and schizophrenia. Nature Rev Drug Discov 3:353–359

Salim S, Sinnarajah S, Kehrl JH, Dessauer CW (2003) Identification of RGS2 and type V adenylyl cyclase interaction sites. J Biol Chem 278:15842–15849

Sanchez-Blazquez P, Rodriguez-Munoz M, Montero C, Garzon (2005) RGS-Rz and RGS9-2 proteins control mu-opioid receptor desensitisation in CNS: the role of activated Galphaz subunits. Neuropharmacology 48:134–150

Siderovski DP, Strockbine B, Behe CI (1999) Whither goest the RGS proteins? Crit Rev Biochem Mol Biol 34:215–251

Snow BE, Betts L, Mangion J, Sondek J, Siderovski DP (1999) Fidelity of G protein beta-subunit association by the G protein gamma-subunit-like domains of RGS6, RGS7, and RGS11. Proc Natl Acad Sci USA 96:6489–6494

Sterne-Marr R, Tesmer JJ, Day PW, Stracquatanio RP, Cilente JA, O'Connor KE, Pronin AN, Benovic JL, Wedegaertner PB (2003) G protein-coupled receptor Kinase 2/G alpha q/11 interaction. A novel surface on a regulator of G protein signaling homology domain for binding G alpha subunits. J Biol Chem 278:6050–6058

Tang KM, Wang GR, Lu P, Karas RH, Aronovitz M, Heximer SP, Kaltenbronn K, Blumer KJ, Siderovski D, Zhu Y, Mendelsohn ME (2003) Regulator of G-protein signaling-2 mediates vascular smooth muscle relaxation and blood pressure. Nat Med 9:1506–1512

Tesmer JJ, Berman DM, Gilman AG, Sprang SR (1997) Structure of RGS4 bound to AlF4–activated G(i alpha1): stabilization of the transition state for GTP hydrolysis. Cell 89:251–261

Traynor JR, Neubig RR (2005) Regulators of G protein signaling and drugs of abuse. Mol Interv 5:30–41

Wang Q, Liu M, Kozasa T, Rothstein JD, Sternweis PC, Neubig RR (2004) Thrombin and lysophosphatidic acid receptors utilize distinct rhoGEFs in prostate cancer cells. J Biol Chem 279:28831–28834

Wang Q, Liu M, Mullah, Siderovski DP, Neubig RR (2002) Receptor-selective effects of endogenous RGS3 and RGS5 to regulate mitogen-activated protein kinase activation in rat vascular smooth muscle cells. J Biol Chem 277:24949–24958

Zachariou V, Georgescu D, Sanchez N, Rahman Z, DiLeone R, Berton O, Neve RL, Sim-Selley LJ, Selley DE, Gold SJ, Nestler EJ (2003) Essential role for RGS9 in opiate action. Proc Natl Acad Sci USA 100:13656–13661

Zeng W, Xu X, Popov S, Mukhopadhyay S, Chidiac P, Swistok J, Danho W, Yagaloff KA, Fisher SL, Ross EM, Muallem S, Wilkie TM (1998) The N-terminal domain of RGS4 confers receptor-selective inhibition of G protein signaling. J Biol Chem 273:34687–34690

Zheng B, Ma YC, Ostrom RS, Lavoie C, Gill GN, Insel PA, Huang XY, Farquhar MG (2001) RGS-PX1, a GAP for GalphaS and sorting nexin in vesicular trafficking. Science 294:1939–1942

Zhong H, Neubig RR (2001) Regulator of G protein signaling proteins: novel multifunctional drug targets. J Pharmacol Exp Ther 297:837–845

Dimeric GPCRs: what did we learn from the metabotropic glutamate receptors?

J.P. Pin[1], *C. Goudet*[1], *J. Kniazeff*[1], *V. Hlavackova*[2], *C. Brock*[1], *V. Binet*[1], *D. Maurel*[1], *P. Rondard*[1], *J. Blahos*[2], and *L. Prezeau*[1]

Summary

For a long time, G-protein coupled receptors (GPCRs) were widely considered to be monomeric entities. However, within the last ten years, increasing amounts of data support the idea that GPCRs can form dimers, either homodimers or heterodimers (Bouvier 2001). Such an observation has raised a number of important issues concerning the activation process: what is the stoichiometry required for the dimer activation? Is one agonist per dimer sufficient? Should both heptahelical domains (HD) be in an active conformation? Here, these issues are examined using the metabotropic glutamate receptors (mGlu receptors), the GPCRs activated by the neurotransmitter glutamate, as a model system. These receptors are constitutive homodimers linked by a disulfide bridge, each subunit being composed of two main domains: a "Venus Flytrap" extracellular domain (VFT) where glutamate binds, and a HD that is common to all GPCRs and is responsible for G-protein activation. Using a strategy allowing perfect control of the subunit composition in a mGlu dimer, we showed that a single agonist per dimer is sufficient for activation but that two agonists are required to reach the full activity. Moreover, thanks to artificial ligands that bind directly into the HDs and stabilize either their active or inactive conformation, we also demonstrated that a single HD per dimer is activated at a time. Such an asymmetric functioning of homodimeric proteins appears surprising. We propose that this phenomenon is also valid for many other GPCRs and results from their association with an asymmetric protein, the heterotrimeric G-protein.

Introduction

Extracellular signals as diverse as light, ions, small molecules like neurotransmitters, peptides or protein hormones act at the surface of their targeted cells, on receptors that transmit this signal inside the cell by activating heterotrimeric G-proteins. These G-protein coupled receptors (GPCRs) represent one of the most important gene families in mammalian genomes and are involved in most physiological processes. As such, it is not surprising that these receptors constitute one of the major targets of the drugs on the market and they are the subjects of intense research in drug development.

[1] Institute of Functional Genomics, Department of Molecular Pharmacology, CNRS UMR5203, Montpellier, France; INSERM U661, Montpellier, France; Université de Montpellier 1, Montpellier, France; Université de Montpellier 2, Montpellier, France
[2] Institute of Molecular Medicine, Prague, Czech Republic

Conn et al.
Insights into Receptor Function
and New Drug Development Targets
© Springer-Verlag Berlin Heidelberg 2006

Since their discovery, these receptors have been considered to be monomeric entities (Chabre and le Maire 2005), but now a large number of studies indicate that these heptahelical integral membrane proteins can form dimers, or even larger oligomers (Angers et al. 2002; Bouvier 2001; George et al. 2002; Milligan 2004). This finding was firmly demonstrated in living cells using energy transfer technologies and helps to explain a number of allosteric properties previously described using ligand binding experiments. Similar findings were obtained with native receptors in their natural environment. Although limited numbers of cases were examined, the data collected suggest that GPCRs oligomerization is a general phenomenon in cell lines as well as in native tissue.

Among the four main classes of GPCRs identified in mammals (Foord et al. 2005), those in class C represent excellent models to examine the precise functioning of GPCR dimers. Actually, all these receptors are known to form constitutive dimers in most of which the two subunits are linked by a disulfide bridge, thus offering a good opportunity to identify the specific role of each subunit in a receptor dimer. Among others, the class C GCPRs are activated by the main neurotransmitters glutamate (the mGlu receptors) and GABA (the $GABA_B$ receptor), Ca^{2+} ions, and sweet and umami taste compounds (Pin et al. 2003).

Eight mGlu receptors have been identified and subclassified into three groups (Conn and Pin 1997). Group I includes mGlu1 and mGlu5, which are post-synaptic receptors coupled to PLC. Groups II (mGlu2 and mGlu3) and III (mGlu4, mGlu6, mGlu7 and mGlu8) are mostly located in presynaptic elements of either glutamatergic or GABAergic synapses and are coupled with Gi proteins, inhibiting adenylyl cyclase and regulating the activity of either K^+ or Ca^{2+} channels (Conn and Pin 1997).

Like any other GPCRs, class C receptors possess a heptahelical domain (HD). However, an additional large extracellular domain called Venus Flytrap (VFT) is present and constitutes the binding site of natural ligands. The structure of the mGlu1 VFT has been solved by X-ray crystallography (Kunishima et al. 2000): it is composed of two opposite lobes surrounding the ligand binding site (Fig. 1). Crystallographic data, as well as functional analysis of mutant receptors, revealed that agonists stabilize a closed form of the VFT, whereas antagonists prevent such closure (Bessis et al. 2002; Kniazeff et al. 2004b; Kunishima et al. 2000; Tsuchiya et al. 2002). Moreover, thanks to data coming from crystallography and FRET studies, a model explaining how the agonists-stabilized-closed VFT lead to G-protein activation by the HD has been proposed (Kubo and Tateyama 2005; Pin et al. 2005). Actually, the crystallographic structures revealed that the agonist binding results in a major movement of one VFT compared to the other, likely producing a relative movement of the HDs within the dimer (Kunishima et al. 2000; Tateyama et al. 2004; Tsuchiya et al. 2002) (Fig. 1). This proposal is consistent with recent FRET data, nicely illustrating that rearrangement in the dimer of HDs is associated with receptor activation (Tateyama et al. 2004).

Such a model for class C dimeric receptor activation helps to explain how the $GABA_B$ receptor could work. This receptor, composed of two distinct but related subunits, $GABA_{B1}$ and $GABA_{B2}$, was the first GPCR identified as an obligatory heterodimer (Jones et al. 1998; Kaupmann et al. 1998; White et al. 1998). Although $GABA_{B1}$ binds all known $GABA_B$ ligands (agonists and competitive antagonists), it is not targeted to the cell surface in the absence of $GABA_{B2}$ (Pagano et al. 2001). In contrast, $GABA_{B2}$ possesses all the required determinants for G-protein activation but does not bind

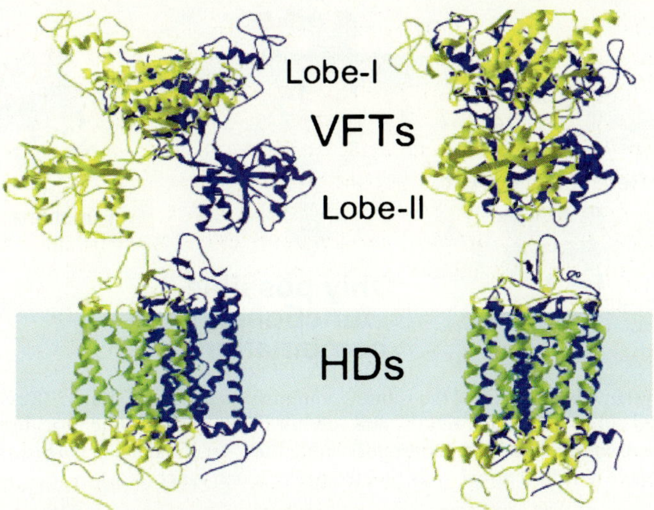

Fig. 1. General structure and activation process of dimeric class C GPCRs. Ribbon view of the crystal structure of the resting Roo (*left*) and fully active Acc (*right*) state of the mGlu1 VFT dimer, and apposition of two rhodopsin structures. The *yellow subunit* is in the *front*; the *blue subunit* is in the *back*. Note the difference in the relative orientation of the two VFTs, probably leading to a different mode of association of the two HDs within the dimer

GABA (Duthey et al. 2002; Galvez et al. 2001; Kniazeff et al. 2002). According to the class C receptor activation model discussed above, agonist binding in the $GABA_{B1}$ VFT results in a relative movement of the two VFTs in the heterodimer, leading to the movement of both of the HDs and thus promoting the activation of the $GABA_{B2}$ HD (Pin et al. 2004). Therefore, specific roles can be assigned to each subunit of the $GABA_B$ receptor: ligand recognition by $GABA_{B1}$, and G-protein activation by $GABA_{B2}$. However, in the case of homodimers like mGlu receptor, does each subunit play a specific role in the activation process? Or are both subunits acting in a symmetrical manner? Answering these questions has been the goal of our workover the last two years.

Results

The quality control system of the $GABA_B$ receptor and its use to control subunit composition in a receptor dimer

To examine the specific role of each subunit in a receptor homodimer, one needs to control precisely the subunit composition within the dimer, such that a single subunit bears a mutation. To achieve that aim, we used the quality control system of the $GABA_B$ receptor. This system is based on the presence of an intracellular retention signal (RSR) in the C-terminal tail of the $GABA_{B1}$ subunit, which is masked upon interaction with the C-terminal tail of the $GABA_{B2}$ subunit (Brock et al. 2005; Margeta-Mitrovic et al. 2000; Pagano et al. 2001). We showed that this system can be transferred to the mGlu receptors by swapping their C-terminal intracellular tails with those of either

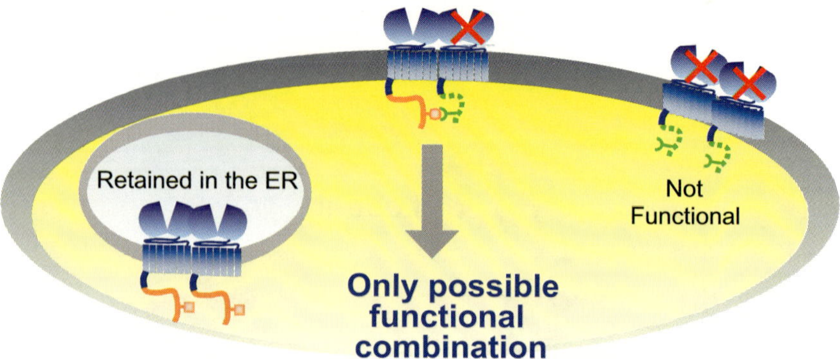

Fig. 2. The system used to control the subunit composition in a mGlu dimer. The C-terminal tail of the GABA$_{B1}$ subunit (*orange*) retains one subunit in the endoplasmic reticulum (ER) unless this subunit is associated with a receptor possessing the C-terminal tail of GABA$_{B2}$ (*green*). If the mGlu-C2 is made non-functional, only the heterodimer can function

GABA$_{B1}$ or GABA$_{B2}$ (Fig. 2). Indeed, mGlu-C1 that possesses the C-terminal tail of GABA$_{B1}$ reaches the cell surface only when associated with mGlu-C2 (Goudet et al. 2005; Hlavackova et al. 2005; Kniazeff et al. 2004a). Interestingly, we recently further improved this system by introducing an intracellular retention sequence in GABA$_{B2}$ such that neither GABA$_{B1}$ nor GABA$_{B2}$ alone is targeted to the surface, and only the heterodimer can escape the intracellular retention (Brock et al. 2005). Such a system allows the analysis of the contribution of each subunit to the dimer activation process by selectively "silencing" a single subunit by mutagenesis.

Functioning of the dimer of VFTs: allostery and symmetry

As a first aim, we decided to examine whether a single agonist could be sufficient to activate a mGlu dimer. We first introduced mutations in the binding site of mGlu5 (Y222A and D304A) that made it unresponsive to agonists (Fig. 3). Then, using the quality control system to express dimers having one wild-type and one mutated binding site, we demonstrated that a single agonist per dimer is sufficient to activate a mGlu dimer (Kniazeff et al. 2004a). However, the response obtained is partial; only when the second site is occupied by a high concentration of agonist can the receptor reach a full activity (Fig. 3). This leads to three important conclusions: 1) a single agonist per dimer is sufficient for activation; 2) two agonists per mGlu dimer are required for full activation; and 3) the binding of the first agonist in one VFT favors the activation of the second one, suggesting a positive allosteric interaction between the two VFTs.

Then, using a receptor dimer composed of one VFT from mGlu5 and one VFT from mGlu2, we analyzed the activity of this receptor with an agonist on one side – closed VFT – and an antagonist on the other – open VFT. Under such conditions, the activity measured was identical to that obtained with a single agonist per dimer, indicating that the partial activity is due to the closure of only a single VFT per dimer (Kniazeff et al. 2004a).

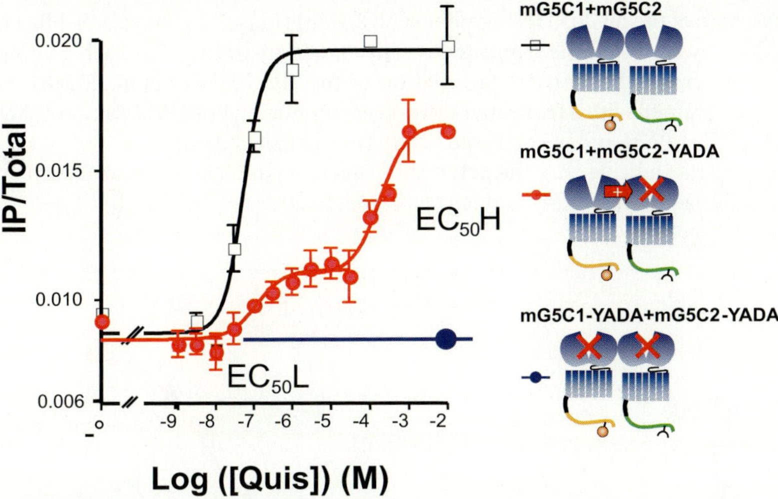

Fig. 3. A single agonist per dimer is sufficient to activate a mGlu dimer, but two are required for full activity. Data show the effect of increasing concentrations of Quisqualate, a mGlu5 agonist, on a mGlu5 dimer possessing two wild-type VFT (*open squares*), or two (*blue circle*) or only one (*red circles*) mutated VFT(s)

These data revealed that the VFT dimer in mGlu receptors functions in a symmetrical manner, with the maximal activity of the receptor dimer requiring both VFTs in a closed active state. But are both HDs in the active state required for full activity of the receptor?

Controlling the conformational state of the HD using allosteric modulators

We next aimed at using a similar approach to examine whether both HDs have to be in an active conformation to activate G-proteins. To do so, we needed to be able to stabilize the HD in either its active or inactive conformation. Thanks to the systematic screening of chemical libraries by pharmaceutical companies, a number of compounds interacting directly in the HD of mGlu receptors have been identified and classified as negative and positive allosteric modulators (NAMs and PAMs, respectively) (Goudet et al. 2004a). NAMs have been shown to exert an inverse agonist activity, therefore stabilizing the HD in a fully inactive conformation. PAMs have no effect on their own on the full-length receptor but facilitate the action of agonists by increasing both their potency and efficacy. We expected that this effect is due to the PAMs binding in agonist-induced activated HDs, thus providing a further stabilization of this active state.

To demonstrate that NAMs and PAMs stabilize the inactive and active conformation of the mGlu HDs, we expressed a mGlu5 mutant deleted of its large extracellular domain. This isolated HD of mGlu5 was expressed correctly, targeted to the cell surface and able to activate PLC, as revealed by the constitutive IP (Inositol Phosphate) formation in cells expressing this protein. Interestingly, and as expected, the mGlu5 NAM MPEP was able to inhibit this constitutive activity, confirming that this compound stabilizes the inactive state of the mGlu5 HD (Goudet et al. 2004b) (Fig. 4). In contrast, DFB, a mGlu5 PAM, directly activated the truncated receptor, demonstrating that such a compound acts by stabilizing the active conformation of the HD (Goudet et al. 2004b). Similar data were obtained with a truncated mGlu1 receptor using Ro01-6128 and BAY36-7620, specific mGlu1 PAM and NAM, respectively (unpublished data).

These data indicate that the active and inactive conformations of the mGlu HDs can be stabilized using PAMs and NAMs, respectively.

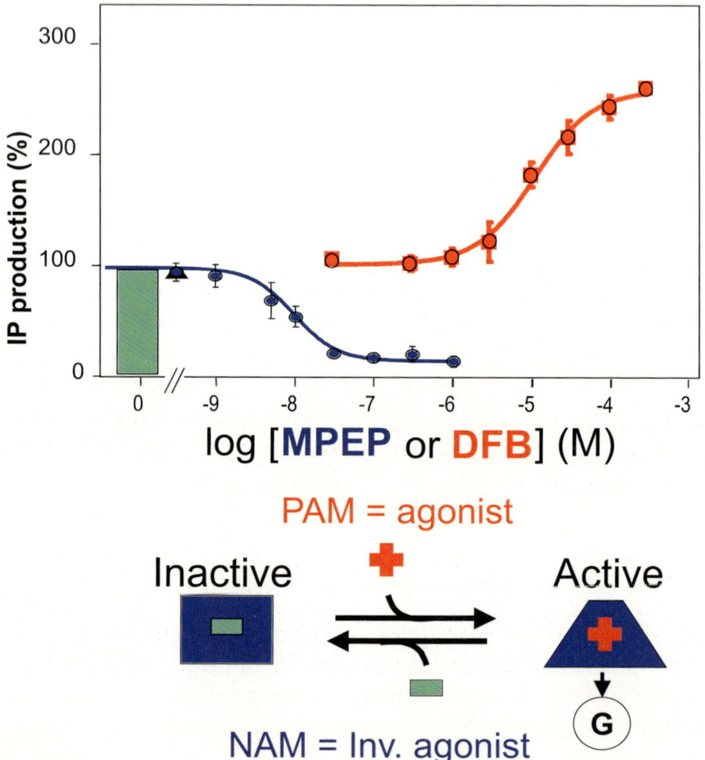

Fig. 4. Controlling the conformation of mGlu HD using NAMs and PAMs. MPEP and DFB (specific mGlu5 NAM and PAM, respectively) stabilize the inactive and active conformations, respectively, of a truncated mGlu5 receptor deleted of its VFT, as illustrated by their inverse (Inv.) agonist and agonist activity, respectively

A single HD per dimer is activated at a time

Using the quality control system, we examined the effect of a NAM acting in a single subunit of a mGlu dimer. To that end, we introduced three point mutations (S668P, C671S and V823A) within the HD of mGlu1 to make it sensitive to the mGlu5 NAM,

MPEP, as previously reported (Pagano et al. 2000). Surprisingly, this MPEP binding in a single subunit had no effect on the dimer activity, indicating that two NAMs per dimer are required to inhibit receptor activity (Hlavackova et al. 2005) (Fig. 5).

Similarly, we examined the effect of a PAM acting in a single subunit. In that case, we used a mGlu5 subunit that was made sensitive to the mGlu1 PAM, Ro01-6128, by introducing three point mutations in the HD (P654S, S657V and L743V), as previously described (Knoflach et al. 2001). We observed that Ro01-6128 had a similar efficacy in increasing agonist potency, whether it binds in one or two subunits in the dimer (Goudet et al. 2005) (Fig. 5). This finding indicates that the binding of a single PAM per dimer is sufficient for the full positive allosteric effect of PAMs.

To explain these surprising data, we proposed that only one of the HD within the dimer can reach an active conformation at a time. Because mGlu receptors are homodimers, then either one of the HDs, but not both, reaches the active state. However, when one HD is maintained in the inactive state with a NAM, every receptor dimer will be active because the second HD can still reach the active state. Similarly, this proposal also nicely explains why a single PAM is sufficient for the full enhancing activity.

To confirm this hypothesis, we repeated the above experiments by adding a muta-tion preventing G-proteins activation in one of the subunits only. If this mutated HD is blocked into an inactive state with a NAM, agonist binding should lead to the activation of the associated HD, leading to an increase in receptor activity. On the other hand, stabilization of the mutated HD in the active conformation with a PAM should prevent the associated HD from ever reaching the active state and thus this single PAM should act as a non-competitive antagonist (Fig. 6). Our experimental data nicely confirm both of these proposals (Goudet et al. 2005; Hlavackova et al. 2005), firmly demonstrating the asymmetric functioning of the dimer of HDs.

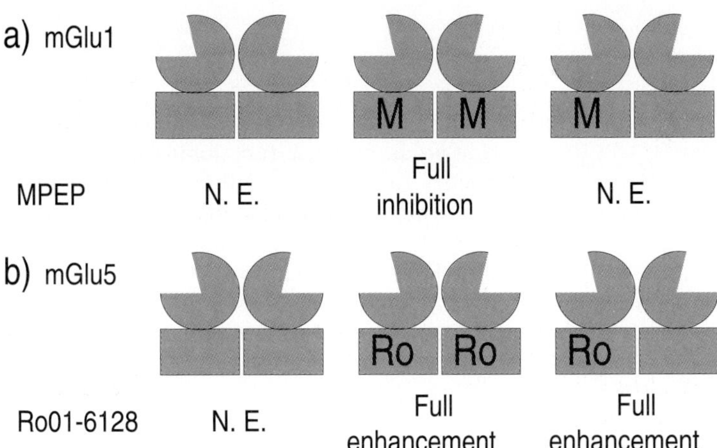

Fig. 5. Two NAMs are required to inhibit mGlu dimer activity, but a single PAM per dimer is sufficient for the full enhancing activity. (a) The effect of the mGlu5-selective NAM, MPEP, was examined on mGlu1 dimer combinations composed of wild-type or MPEP-sensitive mutants (M). (b) The effect of the mGlu1-selective PAM, Ro01-6128, was examined on mGlu5 dimer combinations composed of wild-type or Ro01-6128-sensitive mutants (Ro). N.E., no effect

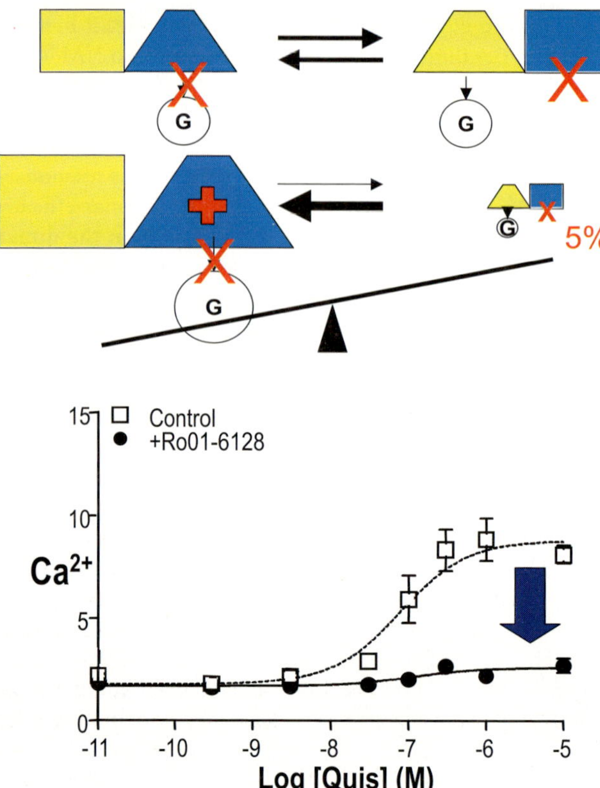

Fig. 6. A single HD per dimer is activated at a time. On the receptor dimer composed of a wild-type mGlu5 HD and a mGlu1 HD mutated in its i3 loop (X), the mGlu1 PAM Ro01-6128 acts as a non-competitive antagonist by stabilizing the active conformation of the G-protein coupling deficient HD, therefore preventing the associated subunit from reaching the active state. The *blue arrow* highlights the non-competitive action of the PAM Ro01-6128 on the receptor dimer combination

Discussion

Our data lead to the surprising finding that a symmetric functioning of the dimer of VFTs in class C GPCRs results in an asymmetric functioning of the dimer of HDs, with only one HD reaching an active conformation within the dimer. But why is the dimer of HDs not working in a symmetrical way in such homodimeric proteins? Although several possibilities can be proposed, the simplest is that an asymmetric interacting protein provides some hindrances that prevent both HDs from reaching their active conformation at the same time. Such an asymmetric protein may well be the heterotrimeric G-protein itself, since it has been proposed to contact both HDs in a GPCR dimer (Baneres and Parello 2003; Filipek et al. 2004).

If only one HD is active at a time, then why is a dimer of class C GPCRs required for function? As mentioned in the introduction, binding of agonists in the VFTs leads to their relative movement, and, as a consequence, to the likely movement of one HD compared to the other. If this is the case, then no signal transduction from the VFT to the HD will be possible in an isolated subunit.

Although distantly related to the HD of mGlu receptors, data indicate that dimeric class A GPCRs also function in an asymmetrical way. First, if one accepts that rhodopsin can form dimers, then, upon excitation with a single photon (our retina can detect a single photon), only one rhodopsin will be activated by trans-retinal, with the associated rhodopsin in the dimer being maintained in an inactive state by cis-retinal, a rhodopsin inverse agonist. Finally, if a single HD can reach an active state in a dimer, there must be only one high-affinity site for an agonist per dimer. Then, binding of one agonist should decrease agonist affinity in the associated subunit. As such, a negative cooperativity in agonist binding should be observed. This situation has indeed recently been clearly demonstrated in a number of receptors, including chemokine (El-Asmar et al. 2005) and glycoprotein hormone receptors (Urizar et al. 2005). But why, for those receptors devoid of a VFT, are two subunits associated in a dimer if only one is active at a time? It is obviously more difficult to answer this question than in the case of the class C GPCRs. One can take this as an argument that a dimer is not required for G-protein activation. However, even though one subunit remains in the inactive state, this does not mean it is not involved in G-protein activation. Indeed, it has been proposed that a relative movement of the beta-gamma dimer relative to the alpha subunit of the G-protein may be involved in the activation. This movement may well result from the interaction of the G-proteins to both protomers of a GPCR dimer, with a single subunit reaching an active state. One may also speculate that, depending on the specific conformation of each subunit, different transduction cascades can be activated. It is now well accepted that GPCRs activate intracellular pathways independently of G-proteins. Maybe in those cases both subunits are required to be in an active state, offering a more complex functioning of these receptors, allowing more subtle regulations for fine tuning of physiological processes.

In conclusion, our study of the functioning of the dimeric mGlu receptor has led us to the surprising discovery of their asymmetric functioning, a property that appears to be shared by other GPCRs, including rhodopsin-like receptors. Such an asymmetric functioning of GPCR dimers may well be related to the proposal that a single G-protein binds to a GPCR dimer, and offers a number of possibilities to fine tune the signal mediated by these receptors.

Acknowledgements. The authors wish to thank Drs. G. Mathis, M. Fink, A. Ansanay and E. Trinquet (CisBio International, Marcoule, France) for constant support and their help in establishing innovative approaches to analyze GPCR dimers. This work was supported by grants from the CNRS, INSERM, Universités de Montpellier 1 & 2, the Action Concertée Incitative "Biologie Cellulaire, Moléculaire et Structurale" of the French Ministry of Research and Technology (grant BCMS328), the European Community (grant LSHB-CT-200-503337), Addex Pharmaceuticals and CisBio International. CB was supported by a FEBS fellowship.

References

Angers S, Salahpour A, and Bouvier M (2002) Dimerization: an emerging concept for G protein-coupled receptor ontogeny and function. Ann Rev Pharmacol Toxicol 42:409–435

Baneres J-L, Parello J (2003) Structure-based analysis of GPCR function. Evidence for a novel pentameric assembly between the dimeric leukotriene B(4) receptor BLT1 and the G-protein. J Mol Biol 329:815–829

Bessis A-S, Rondard P, Gaven F, Brabet I, Triballeau N, Prézeau L, Acher F, Pin J-P (2002) Closure of the Venus Flytrap module of mGlu8 receptor and the activation process: insights from mutations converting antagonists into agonists. Proc Natl Acad Sci USA 99:11097–11102

Table 1. Primary structures of human hypothalamic hormones (including references)

TRH (Burgus et al. 1970; Boler et al. 1969)
pGlu-His-Pro-NH$_2$

GnRH (Matsuo et al. 1971)
pGlu-His-Trp-Ser-Tyr-Gly-Leu-Arg-Pro-Gly-NH$_2$

Somatostatin (SRIF) (Brazeau et al. 1973)
H-Ala-Gly-Cys-Lys-Asn-Phe-Phe-Trp-Lys-Thr-Phe-Thr-Ser-Cys-OH

Somatostatin-28 (SRIF-28) (Pradayrol et al. 1980)
H-Ser-Ala-Asn-Ser-Asn-Pro-Ala-Met-Ala-Pro-Arg-Glu-Arg-Lys-Ala-Gly-Cys-Lys-
Asn-Phe-Phe-Trp-Lys-Thr-Phe-Thr-Ser-Cys-OH

GHRH (Guillemin et al. 1982; Rivier et al. 1982)
H-Tyr-Ala-Asp-Ala-Ile-Phe-Thr-Asn-Ser-Tyr-Arg-Lys-Val-Leu-Gly-Gln-Leu-Ser-Ala-
Arg-Lys-Leu-Leu-Gln-Asp-Ile-Met-Ser-Arg-Gln-Gln-Gly-Glu-Ser-Asn-Gln-Glu-Arg-Gly-
Ala-Arg-Ala-Arg-Leu-NH$_2$

CRF (Rivier et al. 1983)
H-Ser-Glu-Glu-Pro-Pro-Ile-Ser-Leu-Asp-Leu-Thr-Phe-His-Leu-Leu-Arg-Glu-Val-
Leu-Glu-Met-Ala-Arg-Ala-Glu-Gln-Leu-Ala-Gln-Gln-Ala-His-Ser-Asn-Arg-Lys-Leu-Met-
Glu-Ile-Ile-NH$_2$

Urocortin (hUcn) (Donaldson et al. 1996a)
H-Asp-Asn-Pro-Ser-Leu-Ser-Ile-Asp-Leu-Thr-Phe-His-Leu-Leu-Arg-Thr-Leu-Leu-Glu-Leu-
Ala-Arg-Thr-Gln-Ser-Gln-Arg-Glu-Arg-Ala-Glu-Gln-Asn-Arg-Ile-Ile-Phe-Asp-Ser-Val-NH$_2$

Urocortin 2 (hUcn 2) (Reyes et al. 2001)
H-Ile-Val-Leu-Ser-Leu-Asp-Val-Pro-Ile-Gly-Leu-Leu-Gln-Ile-Leu-Leu-Glu-Gln-Ala-Arg-Ala-
Arg-Ala-Ala-Arg-Glu-Gln-Ala-Thr-Thr-Asn-Ala-Arg-Ile-Leu-Ala-Arg-Val-NH$_2$

Urocortin 3 (hUcn 3) (Lewis et al. 2001)
H-Phe-Thr-Leu-Ser-Leu-Asp-Val-Pro-Thr-Asn-Ile-Met-Asn-Leu-Leu-Phe-Asn-Ile-Ala-Lys-
Ala-Lys-Asn-Leu-Arg-Ala-Gln-Ala-Ala-Ala-Asn-Ala-His-Leu-Met-Ala-Gln-Ile-NH$_2$

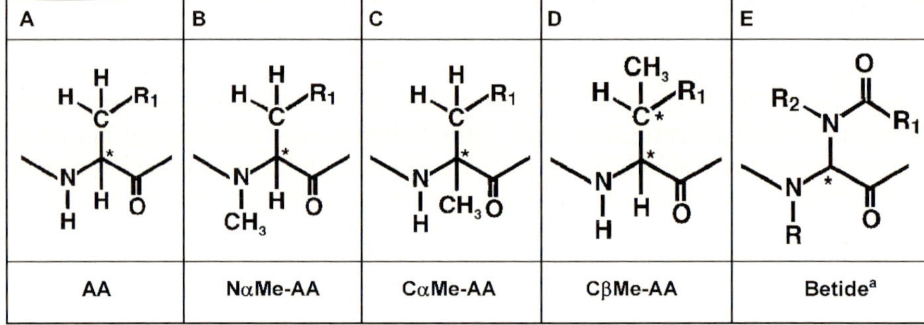

Fig. 1. Peptide chemist's tools: different amino acid scaffolds. * chiral center. ^{a} Betides are acylated and optionally alkylated aminoglycine derivatives. (Rivier et al. 1996)

Gonadotropin releasing hormone (GnRH)

For comparison purposes, we will first review the evolution of the structure of GnRHs over the past 650 million years (MYA). In Table 2, we list the structures of 23 native

GnRHs, nine of which were characterized in tunicates. Structurally, it is noteworthy that all of these structures are ten amino acids long, with the first four and last two residues being identical in all cases but three (a Tyr at position 2 in guinea pig GnRH and a Tyr at position 3 in Lamprey-I and Tunicate-8 GnRHs). Both N-termini (pyroglutamic acid) and amidated C-termini are blocked, suggesting the early existence of the post-translational machinery responsible for these modifications (Fischer and Spiess 1987). It is also noteworthy that mammalian GnRH (found in the sturgeon) dates back to about 450 MYA, whereas Chicken II GnRH, also found in mammals (where it is active), dates back to 530 MYA. Substitutions selected over the years are by natural amino acids, mostly at positions 5–8. It was unexpected to find a Cys residue at position 6 of tunicate-2 GnRH, suggesting that native tunicate-2 GnRH is a dimer.

In summary, evolution introduced amino acid substitutions, dimerization and, either a pair of amino acids amenable to salt bridge formation, such as Asp^5 and Lys^8 in tunicate-1 GnRH, or a glycine at position 6 to induce or allow folding around residues 5–8 to bring about an interaction between N- and C-termini favorable for receptor interaction (Monahan et al. 1973; Kupryszewski et al. 1987).

What did "intelligent design" bring in addition to what nature had contributed? In the case of the design of GnRH superagonists, we can cite the introduction of unnatural amino acids in their D-forms. This design is best illustrated by the structures of the GnRH superagonists that are used therapeutically for sex steroid-dependent patholo-gies, including prostate and breast cancers, precocious puberty, endometriosis and many others. A D-residue at position 6 of the mammalian superagonists is responsible for inducing a beta turn encompassing residues 5–8 and for stabilizing the structure against enzymatic degradation (Monahan et al. 1973). Replacement of the C-terminus glycinamide by an ethylamide (Fujino et al. 1972) also contributes to biological stability, leading to increased potency and long duration of action (Table 3).

The development of GnRH antagonists was much more involved. Indeed, degarelix, the latest, safest and longest acting member of a large family of GnRH antagonists, has only three of the original amino acids found in GnRH at positions 4, 7 and 9 (Jiang et al. 2001). The structures of acyline and degarelix are shown in Fig. 2. Differences between the previous generations of GnRH antagonists, represented by acyline, pertain to the presence of additional opportunities for hydrogen bond formation, resulting from the substitution of the acetyl groups on the side chains of 4-aminophenylalanines at positions 5 and 6 by the L-hydroorotyl and carbamoyl groups, respectively. Whereas both compounds have the same affinity for the GnRH receptor, at a subcutaneous dose of 0.5 mg/rat, acyline in 5% mannitol will inhibit LH and testosterone for about two weeks whereas the same dose of degarelix will achieve inhibition for eight weeks (Jiang et al. 2001; Broqua et al. 2002). From this observation, we conclude that the introduction of urea functions in the design of peptide analogues may be a generally applicable strategy to increase bioavailability and stability in peptide.

Why was so much emphasis put on the development of a GnRH antagonist, in view of the therapeutic success of the superagonists and few reasons to believe that treatment with an antagonist would have distinct advantages? It is noteworthy that agonists and antagonists affect the reproductive system via completely different mechanisms. Whereas treatment with an agonist leads to a reversible desensitization of the GnRH receptor, GnRH antagonists, at least in the first phase of their action, are competing at the receptor for endogenous GnRH, and thus prevent its action. Treatment by GnRH

Table 2. Primary structures and estimated age (in million years, MYA) of native gonadotropin-releasing hormones (including references)

GnRH (Ref.)	1	2	3	4	5	6	7	8	9	10	MYA
Tunicate-1 (Powell et al. 1996)	pGlu	His	Trp	Ser	Asp	Tyr	Phe	Lys	Pro	Gly-NH$_2$	
Tunicate-2 (Powell et al. 1996)	pGlu	His	Trp	Ser	Leu	Cys	His	Ala	Pro	Gly-NH$_2$	
Tunicate-3 (Adams et al. 2003)	pGlu	His	Trp	Ser	Tyr	Glu	Phe	Met	Pro	Gly-NH$_2$	660
Tunicate-4 (Adams et al. 2003)	pGlu	His	Trp	Ser	Asn	Gln	Leu	Thr	Pro	Gly-NH$_2$	
Tunicate-5 (Adams et al. 2003)	pGlu	His	Trp	Ser	Tyr	Glu	Tyr	Met	Pro	Gly-NH$_2$	
Tunicate-6 (Adams et al. 2003)	pGlu	His	Trp	Ser	Lys	Gly	Tyr	Ser	Pro	Gly-NH$_2$	
Tunicate-7 (Adams et al. 2003)	pGlu	His	Trp	Ser	Tyr	Ala	Leu	Ser	Pro	Gly-NH$_2$	
Tunicate-8 (Adams et al. 2003)	pGlu	His	Tyr	Ser	Leu	Ala	Leu	Ser	Pro	Gly-NH$_2$	
Tunicate-9 (Adams et al. 2003)	pGlu	His	Trp	Ser	Asn	Lys	Leu	Ala	Pro	Gly-NH$_2$	
Lamprey-I (Sherwood et al. 1986)	pGlu	His	Tyr	Ser	Leu	Glu	Trp	Lys	Pro	Gly-NH$_2$	560
Lamprey-III (Sower et al. 1993)	pGlu	His	Trp	Ser	His	Asp	Trp	Lys	Pro	Gly-NH$_2$	
Dogfish (Lovejoy et al. 1992)	pGlu	His	Trp	Ser	His	Gly	Trp	Leu	Pro	Gly-NH$_2$	530
Chicken-II (Miyamoto et al. 1984)	pGlu	His	Trp	Ser	His	Gly	Trp	Tyr	Pro	Gly-NH$_2$	
Seabream (Powell et al. 1994)	pGlu	His	Trp	Ser	Tyr	Gly	Leu	Ser	Pro	Gly-NH$_2$	
Pejerrey (Montaner et al. 2001)	pGlu	His	Trp	Ser	Phe	Gly	Leu	Ser	Pro	Gly-NH$_2$	450
Catfish (Ngamvongchon et al. 1992)	pGlu	His	Trp	Ser	His	Gly	Leu	Asn	Pro	Gly-NH$_2$	
Salmon (Sherwood et al. 1983)	pGlu	His	Trp	Ser	Tyr	Gly	Trp	Leu	Pro	Gly-NH$_2$	
Whitefish (Vickers et al. 2004)	pGlu	His	Trp	Ser	Tyr	Gly	Met	Asn	Pro	Gly-NH$_2$	
Herring (Carolsfeld et al. 2000)	pGlu	His	Trp	Ser	His	Gly	Leu	Ser	Pro	Gly-NH$_2$	
Mammal/Sturgeon (Matsuo et al. 1971; Burgus et al. 1972)	pGlu	His	Trp	Ser	Tyr	Gly	Leu	Arg	Pro	Gly-NH$_2$	
Frog (Conlon et al. 1993)	pGlu	His	Trp	Ser	Tyr	Gly	Leu	Trp	Pro	Gly-NH$_2$	360
Chicken-I (Miyamoto et al. 1983)	pGlu	His	Trp	Ser	Tyr	Gly	Leu	Gln	Pro	Gly-NH$_2$	280
Guinea pig (Jimenez-Liñan et al. 1997)	pGlu	Tyr	Trp	Ser	Tyr	Gly	Val	Arg	Pro	Gly-NH$_2$	110

Table 3. Salient GnRH agonist and superagonist structures with relative potencies (including references)

Structure	Relative potency [a]	Refs.
pGlu-His-Trp-Ser-Tyr-Gly-Leu-Arg-Pro-Gly-NH$_2$	1	(Matsuo et al. 1971)
pGlu-His-Trp-Ser-Tyr-Gly-Leu-Arg-Pro-NHCH$_2$CH$_3$	4	(Fujino et al. 1972)
pGlu-His-Trp-Ser-Tyr-DAla-Leu-Arg-Pro-Gly-NH$_2$	4	(Monahan et al. 1973)
pGlu-His-Trp-Ser-Tyr-DTrp-Leu-Arg-Pro-NHCH$_2$CH$_3$	144	(Vale et al. 1977)
pGlu-His-Trp-Ser-Tyr-DHis(B$_2$I)-Leu-Arg-Pro-NHCH$_2$CH$_3$	210	(Vale et al. 1979)

[a] Relative potencies were obtained using an in vitro rat pituitary cell culture assay measuring LH secretion upon incubation with increasing doses of the peptides (Vale et al. 1972)

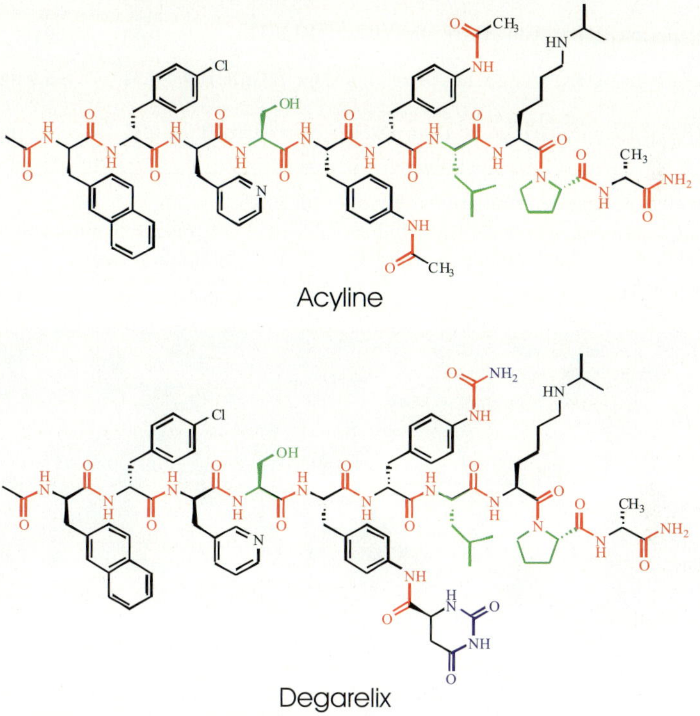

Acyline

Degarelix

Fig. 2. Structures of acyline and degarelix. Differences between the two structures are shown in *blue*. These differences impart significant differences in duration of action after subcutaneous injections (Jiang et al. 2001; Broqua et al. 2002)

antagonists is also reversible. Unique to the treatment by GnRH antagonists is their ability to provide pharmacological castration (as measured by their ability to achieve azoospermia in mammals, including men; Pavlou et al. 1991), whereas treatment with superagonists at best leads to lower sperm count. It is still to be demonstrated that this more profound action is beneficial for the treatment of prostate and breast cancers while being critical for the development of a contraceptive agent.

Significant inroads in the understanding of the interaction of GnRH with its receptor were made using two different and complementary approaches.

1. The design of cyclic and dicyclic GnRH antagonists using covalent bonding of amino acid side chains (Table 4) led to a proposed consensus model of the bioactive conformation of GnRH antagonists determined by NMR (Fig. 3; Koerber et al. 2000). Based on the early observation that all amino acids in the GnRH sequence

Table 4. Structures of potent cyclic GnRH antagonists and their activities

Analog	K_i^a (nM)	AOA [b]	
[Ac-D2Nal^1,DFpa^2,DTrp^3,Asp^4,DArg^6,Dpr^{10}]GnRH	0.36	1.0 µg	(2/10)
[Ac-DAsp^1,DCpa^2,DLys^3,D2Nal^6,DAla^{10}] GnRH	0.82	10 µg	(2/8)
[Ac-D2Nal^1,DCpa^2,DTrp^3,Glu^5,DArg^6,Lys^8,DAla^{10}]GnRH	0.84	50 µg	(0/7)
[Ac-D2Nal^1,DCpa^2,DTrp^3,Asp^4,Glu^5,DArg^6,Lys^8,Dpr^{10}]GnRH	0.32	5.0 µg	(3/20)
[Ac-Asp^1(βAla),DCpa^2,DTrp^3,Asp^4,Dbu^5,D2Nal^6,Dpr^{10}]GnRH	0.15	2.5 µg	(1/8)

[a] K_i are average of three determinations (Rivier et al. 2000).

[b] AOA = antiovulatory assay. Doses and number of rats ovulating over total number of rats treated are shown. The most potent analogs are those that result in the smallest number of rats ovulating at the lowest dose.

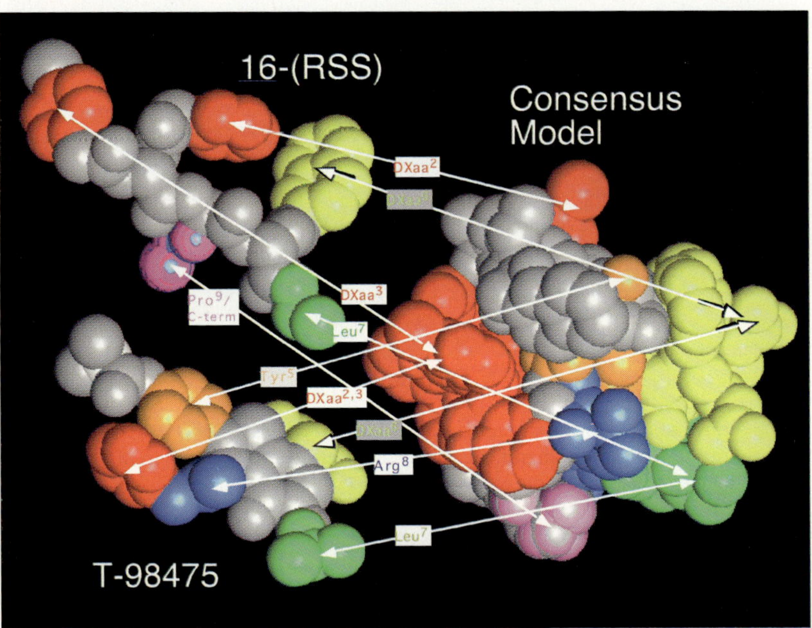

Fig. 3. Consensus pharmacophore of GnRH antagonists (*at right*) with correspondence of functionalities found in 16-(RSS, a tribetide) and T-98475

were critical for receptor activation, the variety of bridges that are compatible with antagonistic biological activity was unexpected, pointing out the critical importance of a peptide backbone's conformation as compared to that of side chain functionality.

Of interest is the correspondence that could be found between that structure and a known nonpeptide ligand, T-98475 (Cho et al. 1998), and a de novo synthesized tribetide, including selected side chains.

2. Extensive site-directed mutagenesis of the receptor led to the identification of the binding pocket and suggested the most likely contact points between receptor and ligand. An excellent recent review covers this topic (Millar et al. 2004).

Somatostatin-14 (SRIF)

At the time of the discovery of the five somatostatin receptors (sst_{1-5}; Meyerhof et al. 1991; Bruno et al. 1992; Yasuda et al. 1992; Rohrer et al. 1993; Xu et al. 1993; Yamada et al. 1992, 1993), analogs of somatostatin had been developed and at least one (octreotide; Bauer et al. 1982) was an established drug for the treatment of acromegaly. In fact, three scaffolds, in addition to the somatostatin ring, had been identified. These are schematically shown in Fig. 4 and include the original mini-somatostatin cyclic (1–8) octapeptide (B; Vale et al. 1977), the cyclic (2–7) octapeptide (C; Bauer et al. 1982) and the equivalent cyclic (2–7) hexapeptide (D; Veber et al. 1979). It soon became obvious that analogs derived from these scaffolds had significant selectivity for sst_2 and, to a lesser extent, for sst_5 and sst_3. Designing somatostatin analogs that would be selective for one or the other of the ssts would have to take these observations into consideration, with a reasonable approach being to impede binding of a poly-potent analog (a pan-somatostatin) to all but one receptor. We will show that this is indeed feasible. It is not our intention to understate the clinical needs for a pan-somatostatin. The recent reports describing SOM-230 (Bruns et al. 2002; Lewis et al. 2003) and KE108 (Reubi et al. 2002) (two poly-potent SRIF analogs) are vivid reminders of the possible divergent needs of the clinic and basic science.

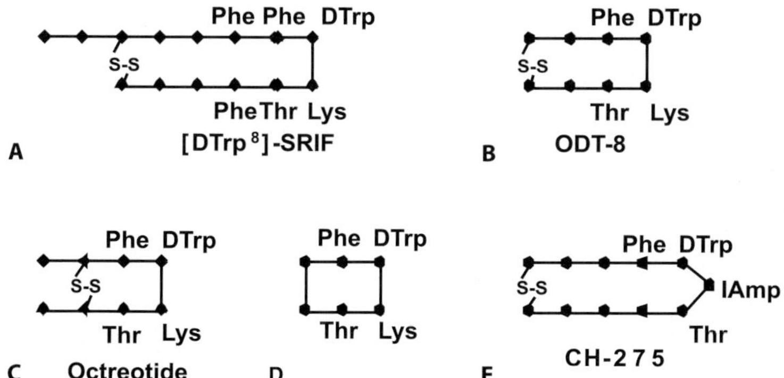

Fig. 4. Illustration of established scaffolds to be used for the design of sst-selective ligands

Sst$_1$-selective SRIF analogs

To illustrate the use of unnatural amino acids, what began as a search for a compatible Lys^9 substitution to increase SRIF's bio-stability led to an sst$_1$-selective analog when 4-aminomethylphenylalanine (Amp^9) in des-AA^{1,2,5}-[DTrp^8,Amp^9]SRIF (Table 5, Compound 3) was substituted by 4[N$'$-(isopropyl)]-aminomethylphenylalanine (IAmp^9) to give des-AA^{1,2,5}-[DTrp^8,IAmp^9]SRIF (Table 5, Compound 4). Substitution of Phe^{11} by Tyr^{11} (Table 5, Compound 5) further improved sst$_1$ affinity, and its iodinated derivative (Table 5, Compound 6) proved to be the desired high affinity label that showed (after ^{125}I-labeling) low or no background after displacement by cold label in autoradiograms (Fig. 5; Rivier et al. 2001a). These analogs are all agonists at sst$_1$.

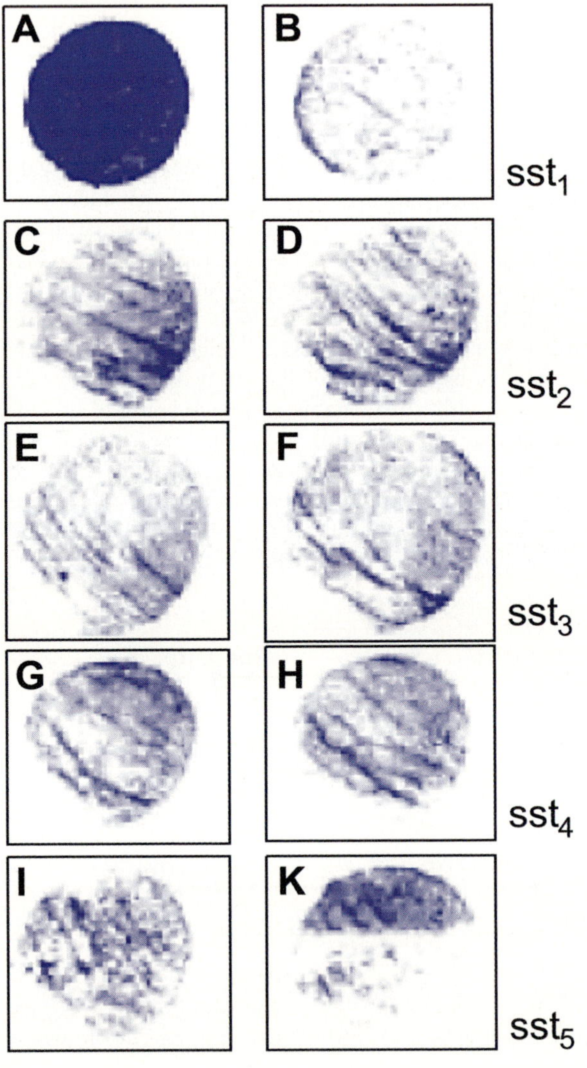

Fig. 5. Binding of ^{125}I-6 to sections of sst$_{1-5}$-expressing cell pellets. Autoradiograms show total binding of ^{125}I-6 to sections of cell pellets expressing sst$_1$ (A), sst$_2$ (C), sst$_3$ (E), sst$_4$ (G) and sst$_5$ (I). Autoradiograms B, D, F, H and K show nonspecific binding of ^{125}I-6 in the presence of 10^{-6} M 6

Table 5. Binding affinities of sst$_1$-selective SRIF analogs (Rivier et al. 2001a)

#	Compound	IC$_{50}$ (nM)				
		sst$_1$	sst$_2$	sst$_3$	sst$_4$	sst$_5$
1	SRIF-28	4.2	2.4	5.3	4.0	4.4
2	Des-AA^{1,2}-[DTrp^8]-SRIF	5.4	0.33	4.3	3.5	16.7
3	Des-AA^{1,2,5}-[DTrp^8,Amp^9]-SRIF	309	213	273	267	190
4	Des-AA^{1,2,5}-[DTrp^8,IAmp^9]-SRIF (CH-275^a)	31	> 10K	345	> 1000	> 10K
5	Des-AA^{1,2,5}-[DTrp^8,IAmp^9,Tyr^{11}]-SRIF	17.1	> 10K	> 1000	> 10K	> 1000
6	Des-AA^{1,2,5}-[DTrp^8,IAmp^9,ITyr^{11}]-SRIF	3.6	> 10K	> 1000	> 1000	> 1000

^{a} CH-275 = H-c[Cys^3-Lys^4-Phe^6-Phe^7-DTrp^8-IAmp^9-Thr^{10}-Phe^{11}-Thr^{12}-Ser^{13}-Cys^{14}]-OH

To illustrate the use of an N$^\alpha$-methyl scan, analogs of des-AA^{1,2,5}-[DTrp/D2Nal^8, IAmp^9]SRIF, with or without a tyrosine or mono-iodotyrosine, were systematically synthesized with the introduction of a backbone N$^\alpha$-methyl group (Fig. 1B) and were tested for binding affinity at the five human somatostatin receptors (sst$_{1-5}$). N$^\alpha$-methylation resulted in loss of affinity at sst$_1$ (2- to > 5-fold) when introduced at residues Lys^4, Phe^6, Phe^7, Thr^{10} and Phe^{11} of the parent compound, des-AA^{1,2,5}-[D2Nal^8,IAmp^9]SRIF. N$^\alpha$-methylation was tolerated at residues Cys^3, D2Nal^8, Thr^{12} and Cys^{14}, with retention of binding affinity and selectivity, and resulted in an increase in binding affinity at positions IAmp^9 and Ser^{13}. In these series, the DTrp^8 substitution versus D2Nal^8 was clearly superior (Erchegyi et al. 2005).

The determination of the conformation of sst$_1$-selective congeners led to a consensus pharmacophore depicted in Fig. 6 (Grace et al. 2005).

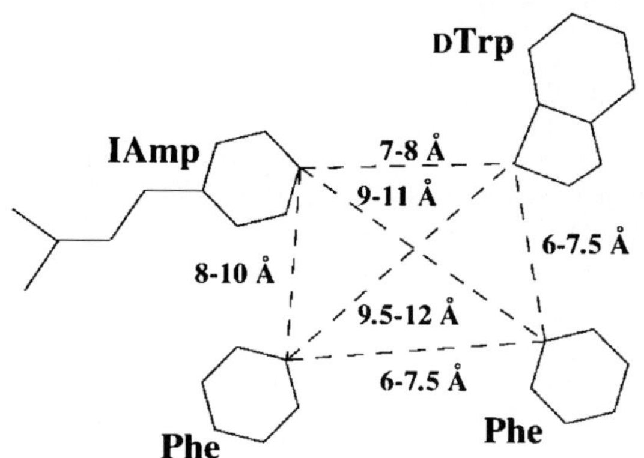

Fig. 6. Schematic drawing of the pharmacophore for sst$_1$-selective analogs (Grace et al. 2005)

Sst$_3$-selective SRIF analogs

To illustrate the use of methyl-betidamino acids (Fig. 1E; R = H, R$_1$ = side chain mimetic; R$_2$ = CH$_3$), we used scaffold B in Fig. 4, where we substituted position 8 with a D-2-naphthylalanine (ODN-8, Compound 7 in Table 6). This analog showed

Table 6. Binding affinities of sst_3-selective SRIF analogs (Reubi et al. 2000)

#	Compound	IC_{50} (nM)				
		sst_1	sst_2	sst_3	sst_4	sst_5
1	SS-28	3.9	3.3	7.1	3.8	3.9
7	Des-AA^{1,2,4,5,12,13}[D2Nal^8]-SRIF (ODN-8)^a	607	173	6.7	56	28
8	Des-AA^{1,2,4,5,12,13}[DAgl^8(Me,2-napht)]-SRIF	> 10K	> 10K	70	> 10K	> 10K
9	Des-AA^{1,2,4,5,12,13}[DCys^3,DAgl^8(Me,2-napht)]-SRIF	> 10K	> 10K	115	> 10K	> 1K
10	Des-AA^{1,2,4,5,12,13}[DCys^3,Tyr^7,DAgl^8(Me,2-napht)]-SRIF	> 10K	> 10K	39	> 10K	> 10K
11	Des-AA^{1,2,4,5,12,13}[DCys^3,Tyr^7,DAgl^8(Me,2-napht)]-Cbm-SRIF (sst$_3$-ODN-8)	> 10K	> 10K	6.7	> 10K	> 10K

^{a} (ODN-8) = H-Cys^3-Phe^6-Phe^7-D2Nal^8-Lys^9-Thr^{10}-Phe^{11}-Cys^{14}-OH

high sst_3-affinity (IC_{50} = 6.7 nM) and moderate sst_3-selectivity (10-fold or more at other receptors). Compound 8, where D2Nal was substituted by its methyl-betidamino acid (DAgl,2-napht), had moderate affinity (IC_{50} = 70 nM) but high sst_3-selectivity (> 100-fold). Additional substitutions (DCys^3 in Table 6, Compound 9 and Tyr^7 in Compound 10) and carbamoylation of the N-terminus yielded sst_3-ODN-8 (Compound 11) with high affinity IC_{50} = 6.7 nM and selectivity (> 100-fold; Reubi et al. 2000). ^{125}I-radioiodinated sst_3-ODN-8 specifically labeled sst_3-expressing cells in autoradiography (Fig. 7A). Sst$_3$-ODN-8 was an antagonist at sst_3. We have no explanation for this observation at this time.

Sst$_4$-selective SRIF analogs

Following the approaches reported above (natural and unnatural amino acid substitutions; Erchegyi et al. 2003b) (Fig. 1A with R_1 variable), betidamino acid scan (Rivier et al. 2003b) (Fig. 1E with R_1 and R_2 variable) and the introduction of beta-methyl amino acids (Erchegyi et al. 2003b), as shown in Fig. 1D (with R_1 variable), high affinity sst_4-selective SRIF analogs were obtained that pointed out the importance of residue 7 (using somatostatin numbering) for sst_4-selectivity. Taking OLT-8 (H-c[Cys-Phe-Phe^7-Trp-Lys-Thr-Phe-Cys]-OH) as the parent analog, we found that substitution of Phe^7 by Ala eliminated all affinity for sst_{1-3} and sst_5 while retaining full affinity at sst_4 (see Table 7). Ultimately, [Tyr^2,Ala^7]OLT-8 that could be ^{125}I-labeled was identified (Erchegyi et al. 2003a,b) (Table 7, Fig. 7B) and a consensus bioactive conformation (Fig. 8A; Grace et al. 2003) similar to that proposed for sst_2/sst_5 and not super-imposable (Fig. 8B; Melacini et al. 1997) was determined using NMR. ^{125}I-labeled [Tyr^2,Ala^7]OLT-8 bound specifically to sst_4 cells (Table 7, Fig. 7B).

These series of experiments confirmed our hypothesis that betidamino acids were a useful tool for on-resin side chain modification of amino acids and that methylated betidamino acid scans were unique in their ability to identify the critical role played

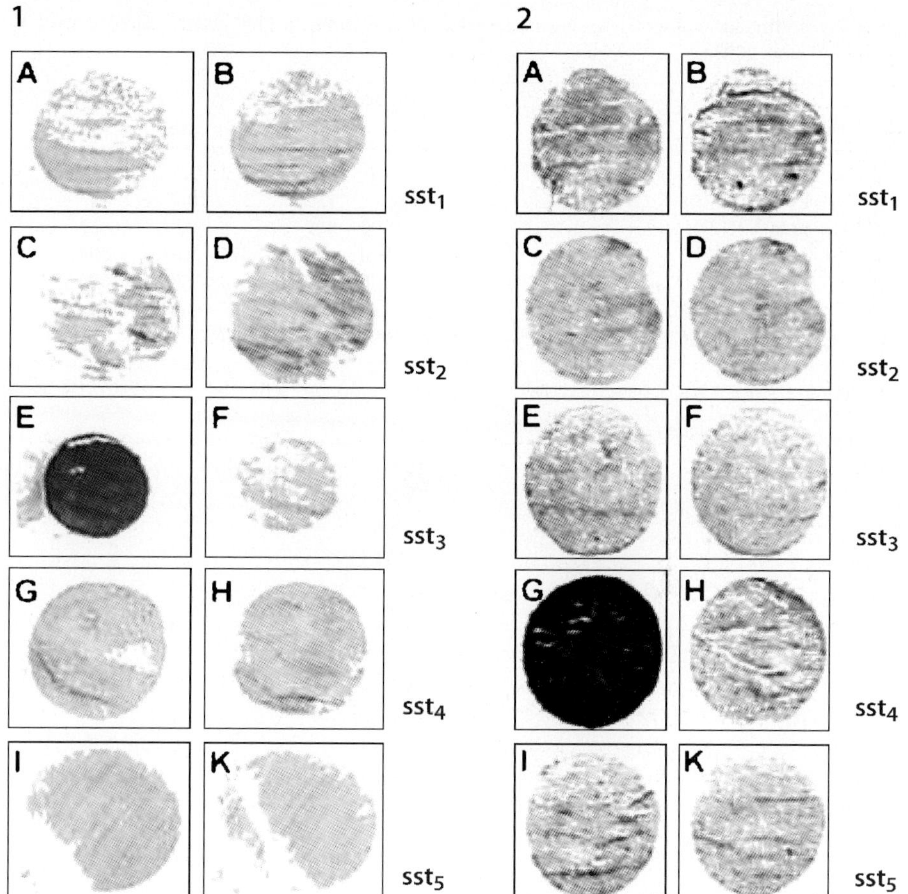

Fig. 7. Binding of ^{125}I-11 (*Panel 1*) and of ^{125}I-14 to sections of sst$_{1-5}$-expressing cell pellets. Autoradiograms show total binding of ^{125}I-11 (*Panel 1*) and ^{125}I-14 (*Panel 2*) to sections of cell pellets expressing sst$_1$ (**A**), sst$_2$ (**C**), sst$_3$ (**E**), sst$_4$ (**G**) and sst$_5$ (**I**). Autoradiograms **B, D, F, H and K** in *Panel 1* show nonspecific binding of ^{125}I-11 (*Panel 1*) and of ^{125}I-14 (*Panel 2*) in the presence of 10^{-6} M **11** and 10^{-6} M **14**, respectively

by side chain orientation during receptor interaction. Having established a broad spectrum of applications, we further proposed that the β-methylated betidamino acids would be functional substitutes for β-methyl amino acids. We now determined that a β-methylated amino acid substitution, in any of the four possible conformers (*R,S*; *R,R*; *S,R*; *S,S*), confered similar biological properties to peptide analogs as the corresponding β-methylated betidamino acid, having only D or L isomers.

To conclude, we confirm our original hypotheses that unusual amino acids such as IAmp can be used in the design of sst$_1$-selective agonists and that constraint of side chains of individual residues using methyl-betidamino acids results in receptor selectivity with maintenance of high affinity in the cases of sst$_3$-selective antagonists and sst$_4$-selective agonists.

Table 7. Binding affinities of sst_4-selective analogs. Critical role of Phe^3 at all receptors but sst_4 (Erchegyi et al. 2003a)

#	Compound [a]	IC$_{50}$ (nM)				
		sst_1	sst_2	sst_3	sst_4	sst_5
1	SRIF-28	3.2	2.3	3.5	2.5	2.4
12	OLT-8	5.3	130	13	0.7	14
13	[Ala^7]-OLT-8	> 1000	807	750	0.84	633
14	[Ala^7]-ODT-8	> 1000	183	897	0.98	199
15	[Tyr^2,Ala^7]-OLT-8	> 1000	622	624	2.0	692
16	[Tyr^2,Ala^7]-ODT-8	330	57	347	1.1	51
17	[ITyr^2,Ala^7]-OLT-8	> 1000	> 1000	1025	3.5	> 1000

[a] OL/DT-8 = H-Cys^3-Phe^6-Phe^7-L/DTrp^8-Lys^9-Thr^{10}-Phe^{11}-Cys^{14}-OH

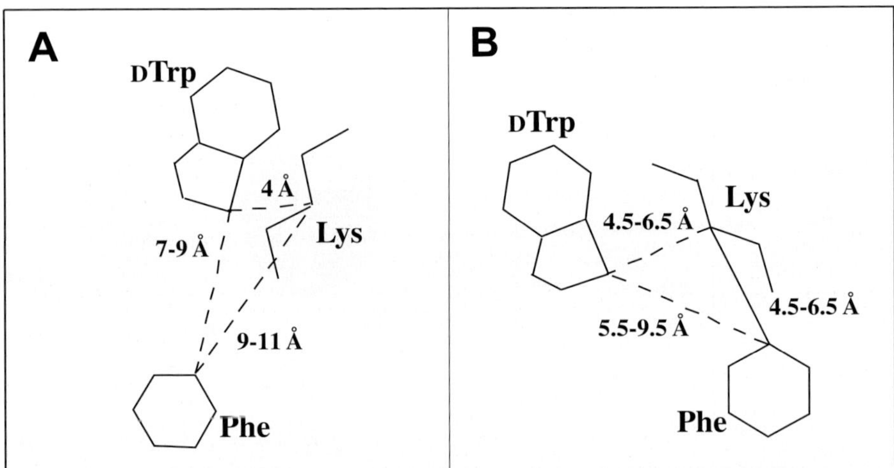

Fig. 8. Schematic drawing of the pharmacophores for sst_4-selective (**A**) and $sst_{2/5}$-selective (**B**) analogs

Corticotropin releasing factor (CRF)

Whereas there is essentially one hormone (GnRH) responsible for reproduction in mammals acting on one receptor, and three hormones (somatostatin-14, somatostatin-24 and cortistatin) responsible for multiple activities mediated by five receptors (sst_{1-5}), there may be as many as four ligands in the CRF family that interact with two families of receptors (CRF_1R and CRF_2R). Ovine CRF (oCRF; Vale et al. 1981) is a 41-residue peptide that stimulates secretion of corticotropin and β-endorphin (Guillemin et al. 1977). The structures of rat CRF (rCRF; Rivier et al. 1983) and human CRF (hCRF; Shibahara et al. 1983) are identical (Table 8).

In 1995, we cloned urocortin (Ucn; Table 8) from rat (rUcn; Vaughan et al. 1995) and human brain (hUcn; Donaldson et al. 1996a,b). Synthetic Ucn is $\sim$ 8 times more potent than CRF at stimulating ACTH release from anterior pituitary cells and 10 times

Table 8. Agonists and antagonists of the CRF family (stressins and astressins)

Compound	Primary sequence (one letter code) [a]	Avg. K_i, nM (binding) CRF_1R	Avg. K_i, nM (binding) $CRF_{2\beta}R$
hCRF	SEEPPISLDLTFHLLREVLEMARAEQ LAQQAHSNRKLMEII-NH_2	0.99 (0.22–4.6)	6.2 (2.0–19)
oCRF	SQEPPISLDLTFHLLREVLEMTKADQ LAQQAHSNRKLLDIA-NH_2	1.2 (0.87–1.6)	52 (21–128)
hUrocortin	DNPSLSIDLTFHLLRTLLEL ARTQSQRERAEQNRIIFDSV-NH_2	0.41 (0.23–0.74)	1.5 (0.89–2.4)
hUrocortin 2	IVLSLDVPIGLLQILLEQAR ARAAREQATTNARILARV-NH_2	> 100	0.50 (0.22–1.16)
hUrocortin 3	FTLSLDVPTNIMNLLFNIAK AKNLRAQAAANAHLMAQI-NH_2	> 100	14 (9.2–20)
α-helical CRF	SQEPPISLDLTFHLLREMLEMAKA EQEAEQAALNRLLLEEA-NH_2	3.5 (1.6–8.0)	1.1 (1.0–1.3)
Stressin$_1$	Ac-PPISLDLTfHLLREVLEXARAEQLAQQ EHSKRKLXEII-NH_2	1.5 (0.9–2.6)	224 (140–370)
Astressin	fHLLREVLEXARAEQLAQ EAHKNRKLXEII-NH_2	0.72 (0.29–1.8)	0.62 (0.49–0.78)
Astressin B	Ac-DLTfHLLREVLEXARAEQZAQ EAHKNRKLXEZI-NH_2	0.56 (0.44–0.74)	1.4 (1.2–1.6)
Astressin$_2$-B	Ac-DLSfHZLRKXIEIEKQEKEKQQA ENNKLLLDZI-NH_2	> 500	1.3 (0.95–1.7)

[a] X = Nle, Z = MeLeu

more potent than CRF at stimulating cAMP accumulation in cells stably expressing CRF_2R (Vaughan et al. 1995; Donaldson et al. 1996a,b). Human CRF and hUcn have only 17 amino acids in common. In 2001, we described the structure of Ucn 2 (Reyes et al. 2001) and Ucn 3 (Lewis et al. 2001) (Table 8), which are N-terminally truncated fragments of stresscopin and stresscopin-related peptides that were independently described by Hsu and Hsueh (2001). Human CRF has only 13 and 12 residues in common with Ucn 2 and Ucn 3, respectively. Ucn 2 and Ucn 3 are ligands with high affinity and selectivity for CRF_2R.

Design of receptor-selective CRF agonists

As mentioned earlier, nature provided us with CRF_2R-selective agonists (Ucn 2 and Ucn 3). Although oCRF is CRF_1R-selective, it still has significant affinity for CRF_2R (Table 8). The recent reports describing two different strategies for inducing CRF_1R selectivity to known structures yielded stressin$_1$ (resulting from side chain to side chain constraints; Rivier et al. 2001b) and cortagine (resulting from chimer formation between sauvagine and hCRF; Tezval et al. 2004) suggest that a long-acting CRF_1R-selective agonist with greater than 300-fold selectivity is within reach.

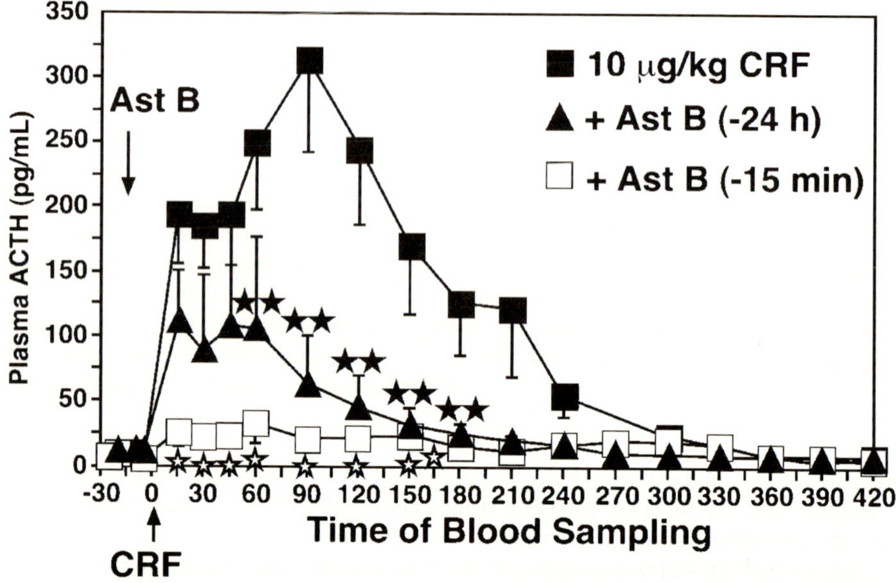

Fig. 11. Profile of astressin B (Ast B) effect on ACTH secretion over time in the rhesus monkey (Broadbear et al. 2004). Plasma ACTH levels after a single injection of 10 µg/kg CRF alone or after 15 min or 24 h pretreatment with 0.1 mg/kg astressin B (* and ** $p < 0.05$ relative to CRF)

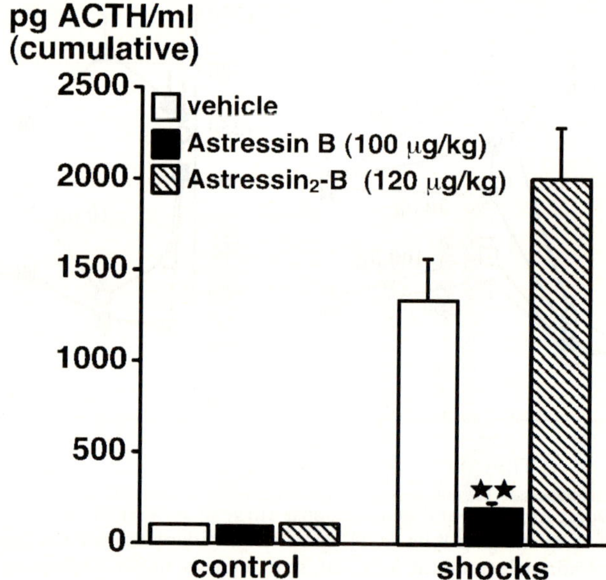

Fig. 12. Comparison between the effect of astressin B and astressin$_2$-B on the ACTH response to shocks over 20 min (Rivier et al. 2003b). Astressin B but not astressin$_2$-B inhibits shock-induced ACTH secretion, demonstrating that the effect is CRF1 mediated. (**, $p < 0.01$ vs. vehicle)

side chain rotational constraints used in the design of SRIF-selective analogs. CRF_2R-selective antagonists (astressin$_2$-B) and CRF_1R receptor-selective agonist (stressin$_1$) result from the use of such strategies. Because CRF receptors are unique and valuable targets for the design of novel drugs addressing gastrointestinal, cardiovascular and CNS illnesses, any progress in the design of novel ligands to these receptors will have considerable clinical significance.

This overview suggests that, given adequate resources, including reliable and quantitative biological in vitro and in vivo tests, strategies that include the development of novel and often constrained amino acids and scaffolds can be identified that will lead to the discovery of specific highly potent ligands to G-protein coupled receptors. The guiding principles applied to the design of GPCR-selective agonists and antagonists presented here are those that led to some desired ligands; they are by no means exclusive. Indeed, we have also investigated substitutions by β-amino acids (Podlech and Seebach 1995) and the introduction of pseudo-bonds, such as reduced amide bonds (Spatola 1983) and peptoids (Simon et al. 1992).

References

Adams BA, Tello J, Erchegyi J, Warby C, Hong DJ, Akinsanya KO, Mackie GO, Vale W, Rivier JE, Sherwood NM (2003) Six novel GnRH hormones are encoded as triplets on each of two genes in protochordate *Ciona intestinalis*. Endocrinology 144:1907–1919

Bauer W, Briner U, Doepfner W, Haller R, Huguenin R, Marbach P, Petcher TJ, Pless J (1982) SMS 201–995: a very potent and selective octapeptide analog of somatostatin with prolonged action. Life Sci 31:1133–1140

Boler J, Enzmann F, Folkers K, Bowers CY, Schally AV (1969) The identity of chemical and hormonal properties of the thyrotropin releasing hormone and pyro-glutamyl-histidyl-proline amide. Biochem Biophys Res Commun 37:705–710

Brazeau P, Vale WW, Burgus R, Ling N, Butcher M, Rivier JE, Guillemin R (1973) Hypothalamic polypeptide that inhibits the secretion of immunoreactive pituitary growth hormone. Science 179:77–79

Broadbear JH, Winger G, Rivier J, Rice KC, Woods JH (2004) Corticotropin-releasing-hormone antagonists, Astressin B and Antalarmin: Differing profiles of activity in Rhesus Monkeys. Neuropsychopharmacology 29:1112–1121

Broqua P, Riviere P, Conn PM, Rivier JE, Aubert MK, Junien J-L (2002) Pharmacological profile of a new, potent and long-acting gonadotropin-releasing hormone antagonist: Degarelix. J Pharmacol Exper Ther 301:95–102

Bruno JF, Xu Y, Song J, Berelowitz M (1992) Molecular cloning and functional expression of a brain-specific somatostatin receptor. Proc Natl Acad Sci USA 89:11151–11155

Bruns C, Lewis I, Briner U, Meno-Tetang G, Weckbecker G (2002) SOM230: a novel somatostatin peptidomimetic with broad somatotropin release inhibiting factor (SRIF) receptor binding and a unique antisecretory profile. Eur J Endocrinol 146:707–716

Burgus R, Dunn T, Desiderio D, Ward D, Vale WW, Guillemin R (1970) Characterization of the ovine hypothalamic hypophysiotropic TSH-releasing factor (TRF). Nature 226:321–325 (25) Guillemin R, Brazeau P, Bohlen P, Esch F, Ling N, Wehrenberg WB (1982) Growth hormone-releasing factor from a human pancreatic tumor that caused acromegaly. Science 218:585–587

Burgus R, Butcher M, Amoss M, Ling N, Monahan M, Rivier J, Fellows R, Blackwell R, Vale WW, Guillemin R (1972) Primary structure of the ovine hypothalamic luteinizing hormone-releasing factor (LRF). Proc Natl Acad Sci USA 69:278–282

Carolsfeld J, Powell JFF, Park M, Fisher WH, Craig AG, Chang JP, Rivier JE, Sherwood NM (2000) Primary structure and function of three gonadotropin-releasing hormones, including a novel form, from an ancient teleost, herring. Endocrinology 141:505–512

Cho N, Harada M, Imaeda T, Imada T, Matsumoto H, Hayase Y, Sasaki S, Furuya S, Suzuki N, Okubo S, Ogi K, Endo S, Onda H, Fujino M (1998) Discovery of a novel, potent, and orally active nonpeptide antagonist of the human luteinizing hormone-releasing hormone (LHRH) receptor. J Med Chem 41:4190–4195

Conlon JM, Collin F, Chiang Y-C, Sower SA, Baudry H (1993) Two molecular forms of gonadotropin-releasing hormone from the brain of the frog, *Rana ridibunda*: purification, characterization, and distribution. Endocrinology 132:2117–2123

Donaldson CJ, Sutton SW, Perrin MH, Corrigan AZ, Lewis KA, Rivier JE, Vaughan JM, Vale WW (1996a) Cloning and characterization of human urocortin. Endocrinology 137:2167–2170

Donaldson CJ, Sutton SW, Perrin MH, Corrigan AZ, Lewis KA, Rivier JE, Vaughan JM, Vale WW (1996b) Cloning and characterization of human urocortin, erratum. Endocrinology 137:3896

Erchegyi J, Waser B, Schaer J-C, Cescato R, Brazeau JF, Rivier J, Reubi JC (2003a) Novel sst_4-selective somatostatin (SRIF) agonists. Part III: Analogues amenable to radiolabeling. J Med Chem 46:5597–5605

Erchegyi J, Penke B, Simon L, Michaelson S, Wenger S, Waser B, Cescato R, Schaer J-C, Reubi JC, Rivier J (2003b) Novel sst_4-selective somatostatin (SRIF) agonists. Part II: Analogues with β-methyl-3-(2-naphthyl)-alanine substitutions at position 8. J Med Chem 46:5587–5596

Erchegyi J, Hoeger CA, Low W, Hoyer D, Waser B, Eltschinger V, Schaer J-C, Cescato R, Reubi JC, Rivier JE (2005) Somatostatin receptor 1 selective analogues: 2. N-methylated scan. J Med Chem 48:507–514

Fischer W, Spiess J (1987) Identification of a mammalian glutaminyl cyclase converting glutaminyl into pyroglutamyl peptides. Proc Natl Acad Sci USA 84:3628–3632

Fisher LA, Rivier C, Rivier J, Brown MR (1991) Differential antagonist activity of α-helical CRF(9–41) in three bioassay systems. Endocrinology 129:1312–1316

Fujino M, Kobayashi S, Obayashi M, Sinagawa S, Fukuda T, Kitada C, Nakayama R, Yamazaki I, White WF, Rippel RH (1972) Structure-activity relationships in the C-terminal part of luteinizing hormone releasing hormone (LH-RH). Biochem Biophys Res Commun 49:863–869

Grace CRR, Erchegyi J, Koerber SC, Reubi JC, Rivier J, Riek R (2003) Novel sst_4-selective somatostatin (SRIF) agonists. Part IV: Three-dimensional consensus structure by NMR. J Med Chem 46:5606–5618

Grace CRR, Durrer L, Koerber SC, Erchegyi J, Reubi JC, Rivier JE, Riek R (2005) Somatostatin receptor 1 selective analogues: 4. Three-dimensional consensus structure by NMR. J Med Chem 48:523–533

Guillemin R, Brazeau P, Bohlen P, Esch F, Ling N, Wehrenberg WB (1982) Growth hormone-releasing factor from a human pancreatic tumor that caused acromegaly. Science 218:585–587

Guillemin R, Vargo T, Rossier J, Minick S, Ling N, Rivier C, Vale WW, Bloom F (1977) β-Endorphin and adrenocorticotropin are secreted concomitantly by the pituitary gland. Science 197:1367–1369

Gulyas J, Rivier C, Perrin M, Koerber SC, Sutton S, Corrigan A, Lahrichi SL, Craig AG, Vale WW, Rivier J (1995) Potent, structurally constrained agonists and competitive antagonists of corticotropin releasing factor (CRF). Proc Natl Acad Sci USA 92:10575–10579

Hsu SY, Hsueh AJ (2001) Human stresscopin and stresscopin-related peptide are selective ligands for the type 2 corticotropin-releasing hormone receptor. Nat Med 7:605–611

Jiang G, Stalewski J, Galyean R, Dykert J, Schteingart C, Broqua P, Aebi A, Aubert ML, Semple G, Robson P, Akinsanya K, Haigh R, Riviere P, Trojnar J, Junien JL, Rivier JE (2001) GnRH antagonists: A new generation of long acting analogues incorporating urea functions at positions 5 and 6. J Med Chem 44:453–467

Jimenez-Liñan M, Rubin BS, King JC (1997) Examination of guinea pig luteinizing hormone-releasing hormone gene reveals a unique decapeptide and existence of two transcripts in the brain. Endocrinology 138:4123–4130

Koerber SC, Rizo J, Struthers RS, Rivier JE (2000) Consensus bioactive conformation of cyclic GnRH antagonists defined by NMR and molecular modeling. J Med Chem 43:819–828

Kornreich WD, Galyean R, Hernandez J-F, Craig AG, Donaldson CJ, Yamamoto G, Rivier C, Vale WW, Rivier JE (1992) Alanine series of ovine corticotropin releasing factor (oCRF): a structure-activity relationship study. J Med Chem 35:1870–1876

Kupryszewski G, Rivier CL, Perrin M, Vale WW, Rivier J (1987) Synthesis of a new cyclic gonadotropin releasing hormone (GnRH) antagonist. Polish J Chem 61:627–630

Lederis K, Letter A, McMaster D, Moore G (1982) Complete amino acid sequence of urotensin I, a hypotensive and corticotropin-releasing neuropeptide from *Catostomus*. Science 218:162–164

Lederis K, McMaster D, Ichikawa T, MacCannell KL, Rivier J, Rivier C, Vale W, Fryer J, Kobayashi Y (1983) Isolation, analysis of structure, synthesis, and biological actions of urotensin I neuropeptides. Can J Biochem 61:602–614

Lewis K, Li C, Perrin MH, Blount A, Kunitake K, Donaldson C, Vaughan J, Reyes TM, Gulyas J, Fischer W, Bilezikjian L, Rivier J, Sawchenko PE, Vale WW (2001) Identification of Urocortin III, an additional member of the corticotropin-releasing factor (CRF) family with high affinity for the CRF2 receptor. Proc Natl Acad Sci USA 98:7570–7575

Lewis I, Bauer W, Albert R, Chandramouli N, Pless J, Weckbecker G, Bruns C (2003) A novel somatostatin mimic with broad somatotropin release inhibitory factor receptor binding and superior therapeutic potential. J Med Chem 46:2334–2344

Lovejoy DA, Fischer WH, Ngamvongchon S, Craig AG, Nahorniak CS, Peter RE, Rivier JE, Sherwood NM (1992) Distinct sequence of gonadotropin-releasing hormone (GnRH) in dogfish brain provides insight into GnRH evolution. Proc Natl Acad Sci USA 89:6373–6377

Matsuo H, Baba Y, Nair RM, Arimura A, Schally AV (1971) Structure of the porcine LH- and FSH-releasing hormone. I. The proposed amino acid sequence. Biochem Biophys Res Commun 43:1334–1339

Melacini G, Zhu Q, Osapay G, Goodman M (1997) A refined model for the somatostatin pharmacophore: Conformational analysis of lanthionine-sandostatin analogs. J Med Chem 40:2252–2258

Meyerhof W, Paust HJ, Schoenrock C, Richter D (1991) Cloning of a cDNA encoding a novel putative G-protein-coupled receptor expressed in specific rat brain regions. DNA Cell Biol 10:689–694

Millar RP, Lu ZL, Pawson AJ, Flanagan CA, Morgan K, Maudsley SR (2004) Gonadotropin-releasing hormone receptors. Endocr Rev 25:235–275

Miranda A, Lahrichi SL, Gulyas J, Koerber SC, Craig AG, Corrigan A, Rivier C, Vale WW, Rivier JE (1997) Constrained corticotropin releasing factor (CRF) antagonists with i-(i+3) Glu–Lys bridges. J Med Chem 40:3651–3658

Miyamoto K, Hasegawa Y, Igarashi M, Chino N, Sakakibara S, Kangawa K, Matsuo H (1983) Evidence that chicken hypothalamic luteinizing hormone-releasing hormone is [Gln[8]]-LH-RH. Life Sci 32:1341–1347

Miyamoto K, Hasegawa Y, Nomura M, Igarashi M, Kangawa K, Matsuo H (1984) Identification of the second gonadotropin-releasing hormone in chicken hypothalamus: evidence that gonadotropin secretion is probably controlled by two distinct gonadotropin-releasing hormones in avian species. Proc Natl Acad Sci USA 81:3874–3878

Monahan M, Amoss M, Anderson H, Vale W (1973) Synthetic analogs of the hypothalamic luteinizing hormone releasing factor with increased agonist or antagonist properties. Biochem 12:4616–4620

Montaner AD, Affanni JM, Park M, Fischer WH, Craig AG, Chang JP, Somoza GM, Rivier JE, Sherwood NM (2001) Primary structure of a novel gonadotropin-releasing hormone (GnRH) variant in the brain of pejerry (*Odontesthes bonariensis*). Endocrinology 142:1453–1460

Montecucchi PC, Henschen A (1981) Amino acid composition and sequence analysis of sauvagine, a new active peptide from skin of *Phyllomedusa sauvagei*. Int J Pept Prot Res 18:113–120

Montecucchi PC, Gozzini L (1982) Secondary structure prediction of sauvagine, a novel biologically active polypeptide from a frog. Int J Pept Prot Res 20:139–143

Murthy ASN, Mains RE, Eipper BA (1986) Purification and characterization of peptidylglycine α-amidating monooxygenes from bovine neurointermediate pituitary. J Biol Chem 261:1815–1822

Ngamvongchon S, Lovejoy DA, Fischer WH, Craig AG, Nahorniak CS, Peter RE, Rivier JE, Sherwood NM (1992) Primary structures of two forms of gonadotropin-releasing hormone, one distinct and one conserved, from catfish brain. Mol Cell Neurosci 3:17–22

Pallai PV, Mabilia M, Goodman M, Vale WW, Rivier JE (1983) Structural homology of corticotropin-releasing factor, sauvagine, and urotensin I: Circular dichroism and prediction studies. Proc Natl Acad Sci USA 80:6770–6774

Pavlou SN, Brewer K, Farley MG, Lindner J, Bastias M-C, Rogers BJ, Rivier JE, Swift LL, Vale WW, Conn PM, Herbert CM (1991) Combined administration of a GnRH antagonist and testosterone in men induces reversible azoospermia without loss of libido. J Clin Endocrinol Metab 73:1360–1369

Podlech J, Seebach D (1995) On the preparation of b-amino acids from a-amino acids using the Arndt-Eistert reaction: Scope, limitations and stereoselectivity. Application to carbohydrate peptidation. Stereoselective α-alkylations of some b-amino acids. Liebigs Ann 1217–1228

Powell JFF, Zohar Y, Elizur A, Park C, Fischer WH, Craig AG, Rivier JE, Lovejoy DA, Sherwood NM (1994) Three forms of gonadotropin-releasing hormone characterized from brains of one species. Proc Natl Acad Sci USA 91:12081–12085

Powell JFF, Reska-Skinner SM, Prakash OM, Fischer WH, Park M, Rivier JE, Craig AG, Mackie GO, Sherwood NM (1996) Two new forms of gonadotropin-releasing hormone in a protochordate and the evolutionary implications. Proc Natl Acad Sci USA 93:10461–10464

Pradayrol L, Jörnvall H, Mutt V, Ribet A (1980) N-terminally extended somatostatin: the primary structure of somatostatin-28. FEBS Lett 109:55–58

Reubi JC, Schaer J-C, Wenger S, Hoeger C, Erchegyi J, Waser B, Rivier J (2000) SST_3-selective potent peptidic somatostatin receptor antagonists. Proc Natl Acad Sci USA 97:13973–13978

Reubi JC, Eisenwiener KP, Rink H, Waser B, Macke HR (2002) A new peptidic somatostatin agonist with high affinity to all five somatostatin receptors. Eur J Pharmacol 456:45–49

Reycs TM, Lewis K, Perrin MH, Kunitake KS, Vaughan J, Arias CA, Hogenesch JB, Gulyas J, Rivier J, Vale WW, Sawchenko PE (2001) Urocortin II: A member of the corticotropin-releasing factor (CRF) neuropeptide family that is selectively bound by type 2 CRF receptors. Proc Natl Acad Sci USA 98:2843–2848

Rijkers DT, Kruijtzer JA, Van Oostenbrugge M, Ronken E, Den Hartog JA, Liskamp RM (2004) Structure-activity studies on the corticotropin releasing factor antagonist astressin, leading to a minimal sequence necessary for antagonistic activity. Chembiochem 5:340–348

Rivier J, Spiess J, Thorner M, Vale W (1982) Characterization of a growth hormone releasing factor from a human pancreatic islet tumor. Nature 300:276–278

Rivier J, Spiess J, Vale W (1983) Characterization of rat hypothalamic corticotropin-releasing factor. Proc Natl Acad Sci USA 80:4851–4855

Rivier J, Rivier C, Vale W (1984) Synthetic competitive antagonists of corticotropin releasing factor: Effect on ACTH secretion in the rat. Science 224:889–891

Rivier J, Rivier C, Galyean R, Miranda A, Miller C, Craig AG, Yamamoto G, Brown M, Vale W (1993) Single point D-substituted corticotropin releasing factor analogues: Effects on potency and physicochemical characteristics. J Med Chem 36:2851–2859

Rivier JE, Jiang G-C, Koerber SC, Porter J, Craig AG, Hoeger C (1996) Betidamino acids: Versatile and constrained scaffolds for drug discovery. Proc Natl Acad Sci USA 93:2031–2036

Rivier J, Gulyas J, Corrigan A, Craig AG, Martinez V, Taché Y, Vale W, Rivier C (1998) Astressin analogues (CRF antagonists) with extended duration of action in the rat. J Med Chem 41:5012–5019

Rivier JE, Struthers RS, Porter J, Lahrichi SL, Jiang G, Cervini LA, Ibea M, Kirby DA, Koerber SC, Rivier CL (2000) Design of potent dicyclic (4–10/5–8) gonadotropin releasing hormone (GnRH) antagonists. J Med Chem 43:784–796

Rivier JE, Hoeger C, Erchegyi J, Gulyas J, DeBoard R, Craig AG, Koerber SC, Wenger S, Waser B, Schaer J-C, Reubi JC (2001a) Potent somatostatin undecapeptide agonists selective for somatostatin receptor 1 (sst1). J Med Chem 44:2238–2246

Rivier JE, Gulyas J, Kirby D, Kunitake K, Donaldson C, Vaughan J, Perrin M, Koerber S, Martinez V, Taché Y, Rivier C, Vale W (2001b) Receptor-selective corticotropin releasing factor analogs. In: 31st Annual Meeting of the Society for Neuroscience: San Diego CA, Session number 413.417

Rivier J, Gulyas J, Kirby D, Low W, Perrin MH, Kunitake K, DiGruccio M, Vaughan J, Reubi JC, Waser B, Koerber SC, Martínez V, Wang LX, Taché Y, Vale W (2002) Potent and long acting corticotropin releasing factor (CRF) receptor 2 selective peptide competitive antagonists. J Med Chem 45:4737–4747

Rivier J, Erchegyi J, Hoeger C, Miller C, Low W, Wenger S, Waser B, Schaer J-C, Reubi JC (2003a) Novel sst4-selective somatostatin (SRIF) agonists. Part I: Lead identification using a betide scan. J Med Chem 46:5579–5586

Rivier C, Grigoriadis D, Rivier J (2003b) Role of corticotropin-releasing factor receptors type 1 and 2 in modulating the rat ACTH response to stressors. Endocrinology 144:2396–2403

Rohrer L, Raulf F, Bruns C, Buettner R, Hofstaedter F, Schule R (1993) Cloning and characterization of a fourth human somatostatin receptor. Proc Natl Acad Sci USA 90:4196–4200

Sherwood N, Eiden L, Brownstein M, Spiess J, Rivier J, Vale W (1983) Characterization of a teleost gonadotropin-releasing hormone. Proc Natl Acad Sci USA 80:2794–2798

Sherwood NM, Sower SA, Marshak DR, Fraser BA, Brownstein MJ (1986) Primary structure of gonadotropin-releasing hormone from lamprey brain. J Biol Chem 261:4812–4819

Shibahara S, Morimoto Y, Furutani Y, Notake M, Takahashi H, Shimizu S, Horikawa S, Numa S (1983) Isolation and sequence analysis of the human corticotropin-releasing factor precursor gene. Embo J 2(5):775–779

Simon RJ, Kania RS, Zuckermann RN, Huebner VD, Jewell DA, Banville S, Ng S, Wang L, Rosenberg S, Marlowe CK, Spellmeyer DC, Tan R, Frankel AD, Santi DV, Cohen FE, Bartlett PA (1992) Peptoids: A modular approach to drug discovery. Proc Natl Acad Sci USA 89:9367–9371

Sower SA, Chiang YC, Lovas S, Conlon JM (1993) Primary structure and biological activity of a third gonadotropin-releasing hormone from lamprey brain. Endocrinology 132:1125–1131

Spatola AF (1983) Peptide backbone modifications: A structure-activity analysis of peptides containing amide bond surrogates, conformationl constraints, and related backbone replacements. In: Weinstein B (ed) Chemistry and biochemistry of amino acids, peptides and proteins. Marcel Dekker, New York, p 267–307

Tezval H, Jahn O, Todorovic C, Sasse A, Eckart K, Spiess J (2004) Cortagine, a specific agonist of corticotropin-releasing factor receptor subtype 1, is anxiogenic and antidepressive in the mouse model. Proc Natl Acad Sci USA 101:9468–9473

Vale W, Grant G, Amoss M, Blackwell R, Guillemin R (1972) Culture of enzymatically dispersed anterior pituitary cells. Functional validation of a method. Endocrinology 91:562–572

Vale W, Rivier C, Brown M, Rivier J (1977) Pharmacology of TRF, LRF and somatostatin. In: Porter JC (ed) Hypothalamic peptide hormones and pituitary regulation: advances in experimental medicine and biology. Plenum Press, New York, p 123–156

Vale W, Rivier C, Perrin M, Rivier J (1979) Synthetic peptides modulating the secretion of gonadotropins and reproductive functions. In: 6th American Peptide Symposium: Georgetown University, Washington, DC

Vale W, Spiess J, Rivier C, Rivier J (1981) Characterization of a 41 residue ovine hypothalamic peptide that stimulates the secretion of corticotropin and β-endorphin. Science 213:1394–1397

Vaughan JM, Donaldson C, Bittencourt J, Perrin MH, Lewis K, Sutton S, Chan R, Turnbull A, Lovejoy D, Rivier C, Rivier J, Sawchenko PE, Vale W (1995) Urocortin, a mammalian neuropeptide related to fish urotensin I and to corticotropin-releasing factor. Nature 378:287–292

Veber DF, Holly FW, Nutt RF, Bergstrand SJ, Brady SF, Hirschmann R, Glitzer MS, Saperstein R (1979) Highly active cyclic and bicyclic somatostatin analogues of reduced ring size. Nature 280:512–514

Vickers ED, Laberge F, Adams BA, Hara TJ, Sherwood NM (2004) Cloning and localization of three forms of gonadotropin-releasing hormone, including the novel whitefish form, in a salmonid, Coregonus clupeaformis. Biol Reprod 70:1136–1146

Xu Y, Song H, Bruno JF, Berelowitz M (1993) Molecular cloning and sequencing of a human somatostatin receptor, hSSTR4. Biochem Biophys Res Commun 193:648–652

Yamada Y, Reisine T, Law SF, Ihara Y, Kubota A, Kagimoto S, Seino M, Seino Y, Bell GI, Seino S (1992) Somatostatin receptors, an expanding gene family: cloning and functional characterization of human SSTR3, a protein coupled to adenylyl cyclase. Mol Endocrinol 6:2136–2142

Yamada Y, Kagimoto S, Kubota A, Yasuda K, Masuda K, Someya Y, Ihara Y, Li Q, Imura H, Seino S, Seino Y (1993) Cloning, functional expression and pharmacological characterization of a fourth (hSSTR4) and fifth (hSSTR5) human somatostatin receptor subtype. Biochem Biophys Res Commun 195:844–852

Yamada Y, Mizutani K, Mizusawa Y, Hantani Y, Tanaka M, Tanaka Y, Tomimoto M, Sugawara M, Imai N, Yamada H, Okajima N, Haruta J (2004) New class of corticotropin-releasing factor (CRF) antagonists: small peptides having high binding affinity for CRF receptor. J Med Chem 47:1075–1078

Yasuda K, Rens-Domiano S, Breder C, Law SF, Saper C, Reisine T, Bell GI (1992) Cloning of a novel somatostatin receptor, SSTR3, coupled to adenylyl cyclase. J Biol Chem 267:20422–20428

Mutations in G proteins and G protein-coupled receptors in human endocrine diseases

Allen Spiegel[1]

Summary

Naturally occurring mutations in the G protein Gs-α subunit and in a number of G protein-coupled receptors (GPCRs) have been identified in human diseases. Such mutations may lead to loss or gain of function of the encoded protein. Study of such naturally occurring, disease-causing mutations offers unique insights into G protein and GPCR structure and function. In general, diseases caused by GPCR loss-of-function mutations are inherited in autosomal recessive fashion, and those caused by gain-of-function mutations are inherited in autosomal dominant fashion. Endocrine gland dysfunction is the most frequently recognized consequence of GPCR mutation. Loss-of-function mutations in GPCRs for various hormones lead to hormone resistance, manifested as hypofunction of the gland expressing the affected GPCR. Conversely, GPCR gain-of-function mutations lead to hormone-independent activation and hyperfunction of the involved gland.

In recent years, our lab has focused on the extracellular calcium-sensing GPCR (CaR) expressed primarily, but not exclusively, in parathyroid glands and kidney. Loss-of-function CaR mutations lead to a form of hyperparathyroidism, an apparent exception to the general pattern described above but in fact reflecting resistance to the normal inhibition of parathyroid hormone (PTH) secretion by the "hormone" agonist, extracellular Ca^{++}. CaR gain-of-function mutations cause autosomal dominant hypocalcemia (ADH), due to activation of the receptor at subphysiologic concentrations of serum Ca^{++}, leading to "inappropriate" inhibition of PTH secretion. Mutations identified in subjects with ADH are missense mutations clustered in discrete regions of the large extracellular domain (ECD) or seven-transmembrane (7TM) domain of this family 3 GPCR. Expression studies of such mutations have allowed us to identify regions in the ECD and 7TM domains that are critical for activation of the receptor by Ca^{++} and by allosteric modulators of the receptor. I will describe our recent work that helps inform the design of novel therapeutics targeting this important GPCR.

Introduction

Just as mutations in genes encoding a variety of enzymes have been identified in the diseases termed "inborn errors of metabolism" by Garrod, mutations in genes encoding

[1] National Institute of Diabetes and Digestive and Kidney Diseases, Bldg. 31, Rm. 9A/52, National Institutes of Health, Bethesda, MD 20892 USA; current address: Albert Einstein College of Medicine, Belfer 312, 1300 Morris Park Avenue, Bronx, New York 10461, USA

Conn et al.
Insights into Receptor Function
and New Drug Development Targets
© Springer-Verlag Berlin Heidelberg 2006

G proteins and G protein-coupled receptors (GPCRs) have been identified in a number of endocrine diseases that may be termed "inborn errors of signal transduction". As illustrated in Fig. 1, inborn errors of metabolism are caused by loss-of-function mutations leading to deficient enzymatic activity, with a corresponding excess of substrate and deficiency of metabolic product. In contrast, inborn errors of signal transduction comprise both loss- and gain-of-function mutations of G proteins and GPCRs. The former manifest as hormone resistance syndromes in which there exists a deficiency of hormone action despite an excess of hormone (resulting from the usual feedback regulation mechanisms). The latter manifest as hormone-independent endocrine hyperfunction. Mutations in G proteins and GPCRs may occur not only as germline mutations leading to inborn errors of signal transduction but also as somatic mutations that may cause more focal phenotypes in adults. For germline mutations, the particular phenotype caused by a given mutation will be a function of the range of expression of the involved gene, with genes that are more widely expressed leading to a more pleiotropic phenotype. For somatic mutations, focal manifestations may result even from mutation in a ubiquitously expressed gene.

Mutations in G proteins and GPCRs may impair function at any of several steps in the GTPase cycle (Fig. 2). Naturally occurring, germline loss-of-function mutations in the gene encoding the α subunit of the ubiquitously expressed G protein, Gs, coupling many GPCRs to stimulation of cAMP formation, cause the pleiotropic manifestations of the archetypical hormone resistance disorder, pseudohypoparathyroidism. Somatic gain-of-function mutations of the same gene occurring early in development cause McCune-Albright syndrome with endocrine, skin and bone manifestations, whereas somatic mutations of the gene occurring later in life cause more focal manifestations, such as somatotroph pituitary tumors. A more detailed description of the complex regulation of the imprinted Gs-α gene and the disorders resulting from mutations in the gene can be found in a recent review (Spiegel and Weinstein 2004).

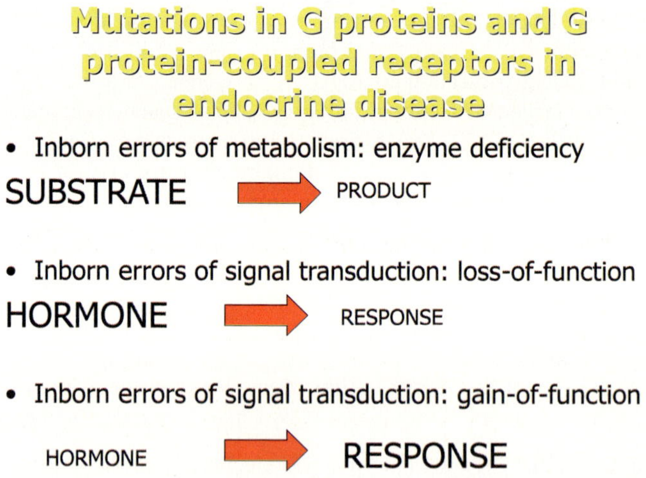

Fig. 1. Inborn errors of metabolism and signal transduction compared (see text)

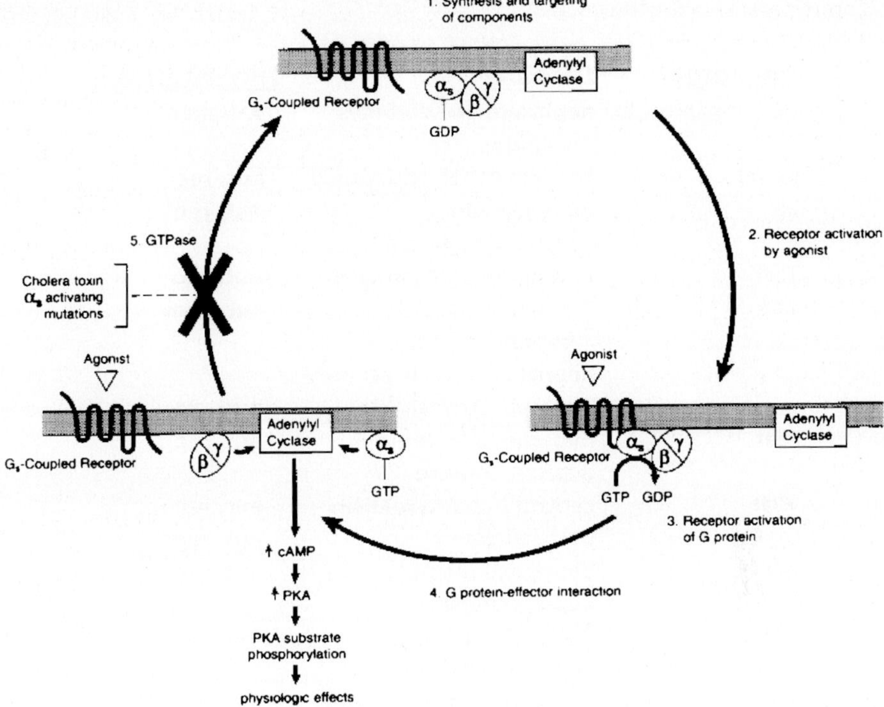

Fig. 2. The G protein GTPase cycle. Potential sites for disease-causing abnormalities are numbered. In each panel, the stippled region denotes the plasma membrane with extracellular above and intracellular below. Under physiologic conditions, effector regulation by G protein subunits is transient and is terminated by the GTPase activity of the α subunit. The latter converts bound GTP to GDP, thus returning the α subunit to its inactivated state with high affinity for the βγ dimer, which reassociates to form the heterotrimer. The figure shows the G protein Gs with its effector, adenylyl cyclase. Activation of adenylyl cyclase generates the intracellular second messenger, cAMP, which activates protein kinase A (PKA). The latter enzyme phosphorylates a variety of proteins that mediate the physiologic effects of agonists for Gs-coupled receptors. Cholera toxin covalently modifies the Gs α subunit, blocking its GTPase activity. Somatic mutations of the Gs α subunit likewise block GTPase activity. In both cases, constitutive activation and agonist-independent cAMP formation result

Endocrine diseases caused by GPCR gene loss-of-function mutations

Clinically significant impairment of signal transduction generally requires loss of function of both alleles of a GPCR gene; thus, most such diseases are autosomal recessive, but there are several exceptions (Fig. 3). Loss-of-function mutations may be missense as well as nonsense or frameshift mutations that truncate the normal receptor protein. They may involve any portion of the receptor, although the membrane-spanning helices are a particularly frequent site. Loss of-function mutations of receptors for ACTH, TSH, FSH, and the hypothalamic hormones – gonadotropin-releasing hormone (GnRH), thyrotropin-releasing hormone (TRH), and growth hormone-releasing

Endocrine diseases caused by GPCR loss of function mutations

Receptor	Disease	Inheritance
V2 vasopressin	nephrogenic diabetes insipidus	X-linked
ACTH	familial ACTH resistance	aut. rec
LH	male pseudo-hermaphroditism	aut. rec
TSH	familial hypothyroidism	aut. rec
CaR	familial hypocalciuric hypercalcemia/	aut. dom
	neonatal severe primary hyperparathyroidism	aut. rec
FSH	hypergonadotropic ovarian failure	aut. rec
TRH	central hypothyroidism	aut. rec
GHRH	GH deficiency	aut. rec
GNRH	hypogonadotropic hypogonadism	aut. rec
PTH	Blomstrand chondrodysplasia	aut. rec
MCR4	obesity	aut. dom

Fig. 3. Endocrine diseases caused by GPCR loss-of-function mutations. The figure lists the receptor gene affected and the typical mode of inheritance of germline mutations. Aut. rec., autosomal recessive; aut. dom, autosomal dominant

hormone (GHRH) – mimic the deficiency of the respective hormones. Subjects with heterozygous loss-of-function mutations of the TSH receptor gene are generally euthyroid with compensatory elevated serum TSH, but homozygous mutations result in congenital hypothyroidism associated with a hypoplastic or even absent thyroid gland. Loss-of-function mutations in LH and PTH/PTHrP receptors cause developmental anomalies, reflecting the critical role of the respective hormones in normal development. Loss-of-function mutations of both copies of the LH receptor gene cause a rare form of 46, XY male pseudohermaphroditism known as Leydig cell hypoplasia. Absence of functional PTH/PTHrP receptors causes a rare, lethal form of dwarfism known as Blomstrand chondrodysplasia. X-linked nephrogenic diabetes insipidus (renal vasopressin resistance) is caused by loss-of-function mutations in the V2 vasopressin receptor gene located on the X chromosome. Males inheriting a mutant gene develop the disease, whereas most females do not show overt disease because random X inactivation results, on average, in 50% normal receptor genes. Identification of the mutation in carrier females facilitates early treatment of affected male neonates to avoid hypernatremia and brain damage. Loss-of-function mutations in the gene encoding the melanocortin 4 receptor, which regulates hypothalamic pathways controlling appetite and energy metabolism, result in a distinct obesity syndrome characterized by hyperphagia and increased linear growth. Inheritance is codominant, with homozygotes showing a more severe phenotype than heterozygotes.

Endocrine diseases caused by GPCR gene gain-of-function mutations

Given the dominant nature of activating mutations, most diseases caused by GPCR gain-of-function mutations are inherited in an autosomal dominant manner (Fig. 4). Unlike loss-of-function mutations, GPCR gain-of-function mutations are almost always missense mutations. Activating missense mutations are thought to disrupt normal inhibitory constraints that maintain the receptor in its inactive conformation. Mutations disrupting these constraints mimic the effects of agonist binding and shift the equilibrium toward the activated state of the receptor.

Germline gain-of-function mutations in the LH and TSH receptor genes may mimic states of hormone excess, familial male precocious puberty, and familial nonautoimmune hyperthyroidism, respectively. Women inheriting gain-of-function mutations in the LH receptor gene do not show precocious puberty because, unlike in males, the combined action of LH and FSH is required for female pubertal development.

Endocrine diseases caused by GPCR gain of function mutations

Receptor	Disease	Inheritance
LH	familial male precocious puberty	aut. dom.
LH	sporadic Leydig cell tumors	somatic
TSH	sporadic hyperfunctional thyroid nodules	somatic
TSH	familial nonautoimmune hyperthyroidism	aut. dom.
CaR	familial hypoparathyroidism	aut. dom.
PTH/PTHrP	Jansen metaphyseal chondrodysplasia	aut. dom.
V2 vasopressin	nephrogenic inappropriate antidiuresis	aut. dom.

Fig. 4. Endocrine diseases caused by GPCR gain-of-function mutations. The figure lists the receptor gene affected and the typical mode of inheritance of germline mutations. Note that somatic mutations may cause sporadic thyroid and testicular Leydig cell tumors. Aut. dom, autosomal dominant

As with activating Gs α mutations, increased cAMP in many endocrine cells leads to increased proliferation and hormone hypersecretion. Thus, somatic gain-of-function mutations of the LH and TSH receptor genes cause sporadic tumors of Leydig cells and the thyroid cells, respectively. Activating mutations of the PTH/PTHrP receptor gene cause Jansen's metaphyseal chondrodysplasia. The phenotype includes hypercalcemia and hypophosphatemia mimicking the effects of PTH hypersecretion, but also abnormal bone development (short-limb dwarfism), reflecting the critical role of PTHrP in endochondral bone formation. Activating mutations of the V2 vasopressin receptor were identified in neonates manifesting a syndrome of inappropriate antidiuresis but lacking the elevated serum vasopressin typically associated with this syndrome.

Overview of the Extracellular calcium-sensing receptor

The cloning of the extracellular Ca^{2+}-sensing receptor (CaR) provides a new paradigm in signal transduction in which an extracellular ion, Ca^{2+}, serves as an agonist for a cell-surface receptor (Brown et al. 1993). The CaR is expressed abundantly in parathyroid and kidney, where its activation inhibits parathyroid hormone (PTH) secretion and promotes urinary Ca^{2+} excretion, respectively (Brown and MacLeod 2001). The CaR is expressed in other tissues, where it might have roles beyond extracellular Ca^{2+} homeostasis (see Chattopadhyay et al. 1998 for a review).

Notwithstanding its unique agonist, the CaR is a member of the GPCR family 3 or C (Bockaert and Pin 1999). All GPCRs share the signature 7TM-spanning domain. The assumption is that GPCR activation involves a conformational change of the membrane-spanning α helices, altering the disposition of intracellular loops, and thereby promoting activation of G proteins. For rhodopsin, a member of GPCR family 1, the three-dimensional (3D) structure of the receptor with its covalently bound ligand, retinal, has been discovered, providing direct evidence for the interaction of ligand with specific residues of the membrane-spanning helices (Palczewski et al. 2000). For members of GPCR family 3, which include, in addition to the CaR, multiple subtypes of metabotropic glutamate receptor (mGluR), the GABA-B receptor and certain taste and pheromone receptors, evidence indicates that agonists bind to a dimeric, Venus flytrap-like (VFT) domain within the large N-terminal ECD of the receptor. The VFT domain is linked to the 7TM domain by a cysteine-rich domain (Fig. 5). Understanding how agonist binding to the VFT domain leads to receptor activation has important implications for designing drugs targeting family 3 GPCRs.

The human CaR (hCaR) is a 1,078-amino acid polypeptide comprising an N-terminal ECD, the 7TM domain and intracellular C-terminus (Fig. 6; see Hu and Spiegel 2003 for review). The ECD contains 11 potential N-linked glycosylation sites of which at least three must be glycosylated for cell surface expression. Ca^{2+} activates the CaR at millimolar concentrations, implying a much lower affinity Ca^{2+}-binding site than for intracellular Ca^{2+}-binding proteins such as calmodulin.

The discovery of the 3D structure of the VFT domain of the rat mGluR1 (Kunishima et al. 2000) offers important insights into agonist-promoted conformational changes, which are probably relevant for the CaR and other members of family 3. The crystal structure of the glutamate-bound form of the mGluR1 VFT revealed the key residues in

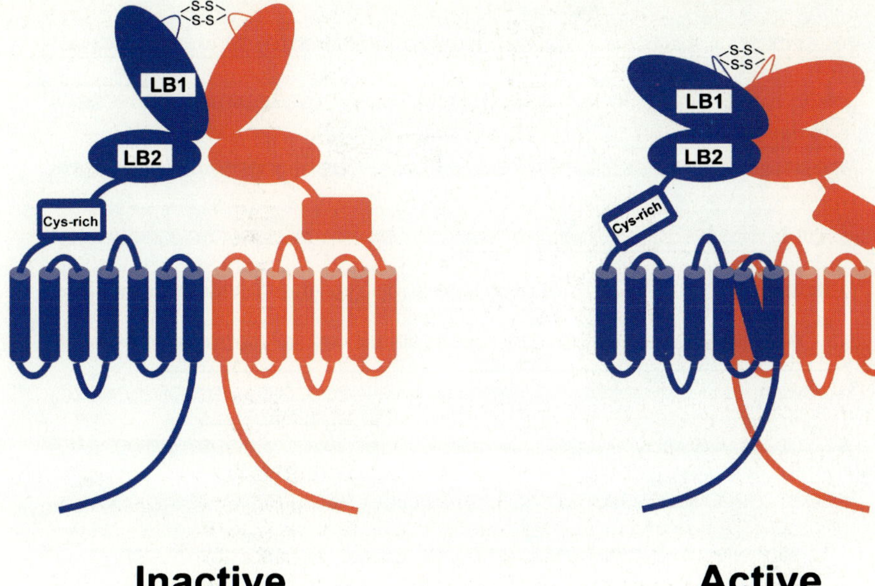

Fig. 5. The CaR is shown schematically in its inactive (free) form and its active (agonist-bound) form. Protomers of the CaR dimer are *colored blue* and *red*, respectively. The VFT domain (LB1 and LB2) and cysteine-rich domain (Cys-rich) of one protomer are labeled, and loop 2 with its two intermolecular disulfide bonds linking each LB1 protomer is shown. The 7TM domain is shown with its three extracellular loops (*top*) and its three intracellular loops and C terminus (*bottom*) connecting seven membrane-spanning α helices (*cylinders*). Agonist binding (not shown) to a cleft between LB1 and LB2 leads to VFT closure and a rotation about the dimer interface (note the change in loop 2 configuration)

LB1 and LB2 involved in agonist binding. Studies of chimeric receptors show that the predominant agonist-binding site for the CaR, and probably most other family 3 GPCRs, resides within the VFT domain. The specific amino acids responsible for Ca^{2+} binding to the CaR have not been definitively identified, but three residues, Ser147, Ser170 and Asp190, corresponding to amino acids in the mGluR1 glutamate binding site, when artificially mutated to alanine, impair CaR activation. L-Amino acids allosterically enhance CaR sensitivity to Ca^{2+}, and studies of the Ser170Ala mutant suggest that the amino acid-binding site is related to that for Ca^{2+} itself.

The CaR is a homodimer linked by intermolecular disulfides at cysteines 129 and 131, as well as by noncovalent interactions along a dimer interface involving both lobes 1 and 2 of the VFT domain (Fig. 5). Comparison of the glutamate-bound, "active" versus antagonist-bound, "inactive" structures of the mGluR1 VFT revealed several important differences (8): 1) the VFT is closed in the glutamate-bound and open in the antagonist-bound structures; 2) residues equivalent to hCaR 117–123 in loop 2 form an ordered extension of an α helix of LB1 in the inactive form, but are disordered, along with the remainder of loop 2, in the active form; 3) agonist-promoted VFT closure leads to a 70° rotation of one monomer relative to the other about an axis perpendicular to the dimer interface; and 4) VFT closure-promoted rotation of the monomers permits LB2

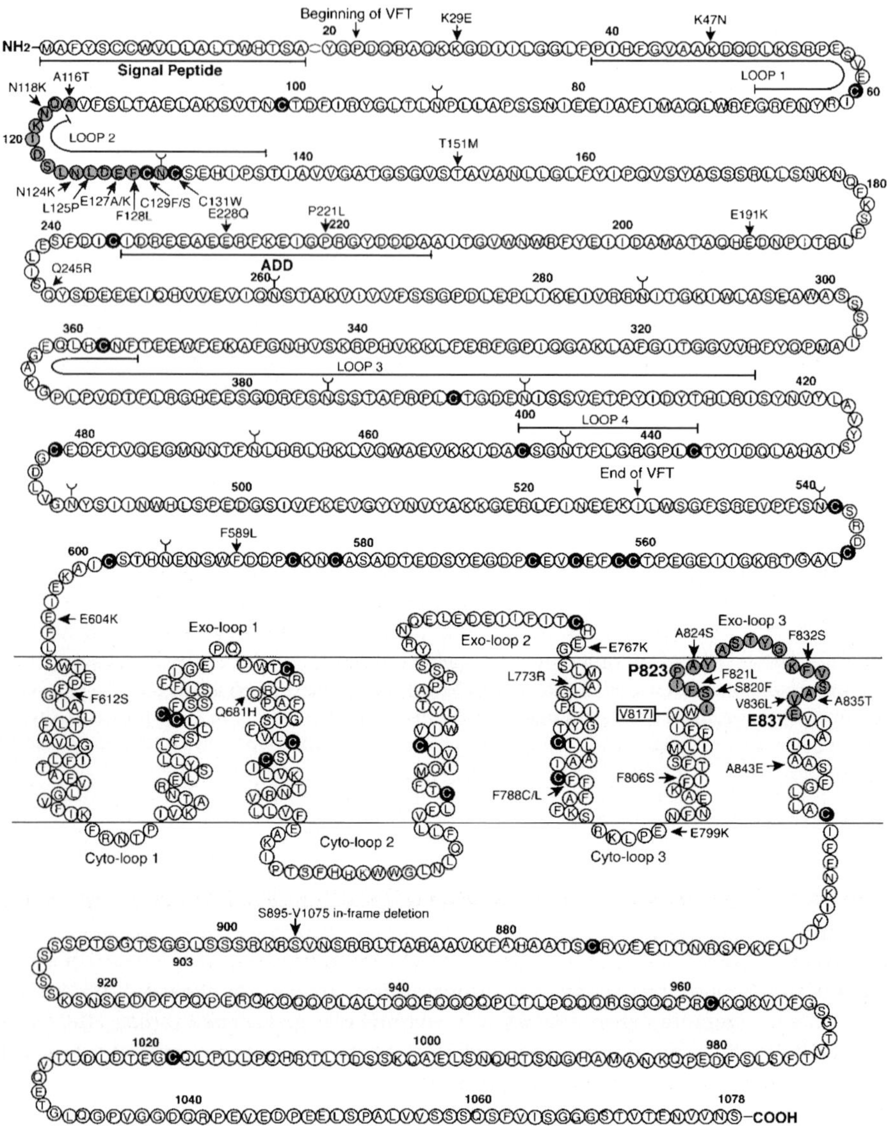

Fig. 6. Schematic diagram showing the amino acid sequence of the hCaR with boundaries of transmembrane helices based on alignment with rhodopsin. The location of signal peptide, *N*-linked glycosylation sites and the sequence of synthetic polypeptide used to raise monoclonal antibody ADD are indicated. All cysteines are shown in *black background*. The beginning and end of the VFT domain and the four loops in lobe 1 of the VFT are indicated. Naturally occurring activating mutations identified in the hCaR, as well as the inactivating V817I mutation (*boxed*), are indicated. Glu837, shown to be involved in binding of the allosteric modulators NPS R-568 and NPS 2143, and Pro823, reported to be critical for the function of the receptor, are shown in *bold print*. The two regions with clustering ADH mutations, residues 116–131 and residues 819–837, are *shaded*

domains to move 26A° closer than in the open VFT conformation, where electrostatic repulsion keeps them further apart. Apposition of the LB2 domains in the agonist-bound state might cause concomitant movement of the cysteine-rich domains linked to LB2. These changes in the CaR are shown in Fig. 5.

The VFT and 7TM domains are linked by an 84-residue region containing nine closely spaced cysteines (Fig. 6), termed the cysteine-rich domain. With the exception of the GABA-B receptor, which lacks this domain, other family 3 GPCRs contain the same nine cysteines with conserved spacing. Mutation of any of these cysteines to serine severely impairs the expression and function of the CaR. Although chimeric hCaRs, in which the mGluR1 cysteine-rich domain is substituted for that of the hCaR, preserve some degree of function, deletion of the cysteine-rich domain abolishes CaR activation, in spite of the preservation of some cell-surface expression. This finding suggests that the cysteine-rich domain plays a key role in signal transmission between the VFT and 7TM domains.

Analysis of the products of tobacco etch virus protease cleavage at a site artificially inserted between the VFT and cysteine-rich domains demonstrated that the dimeric CaR VFT domain is not linked by disulfide bonds to either the cysteine-rich or 7TM domains.

A truncation mutant with N-terminal residues 1–20 of bovine rhodopsin fused to hCaR Ala600 (Rho-C–hCaR) shows excellent cell-surface expression and is activated by Ca^{2+} when added with an allosteric modulator, NPS R-568. These results suggest that the 7TM domain, in addition to the VFT, might contain sites for polycation binding and CaR activation. Mutagenesis of the acidic residues in extracellular loops 1–3, however, does not abolish Ca^{2+} activation of the receptor. Much of the 216 residue C-terminus of the receptor (residues 889–1078) can be truncated without impairing cell-surface expression and activation. Nonetheless, the C-terminus might be responsible for other properties of the CaR, such as binding to a scaffold protein, filamin-A.

Diseases Caused by Loss- and Gain-of-function Mutations of the CaR

The importance of the CaR in extracellular Ca^{2+} homeostasis is underscored by the identification of inactivating mutations in the CaR gene as the cause of familial hypocalciuric hypercalcemia (FHH) and neonatal severe primary hyperparathyroidisim (NSPHT) and the identification of activating mutations as the cause of autosomal dominant hypocalcemia/hypoparathyroidism (ADH).

Inactivating mutations of the CaR cause a right-shift in set point for Ca^{2+} inhibition of PTH secretion (Fig. 7) and for stimulation of urinary Ca^{2+} excretion, leading to relative hypercalcemia and hypocalciuria in subjects with FHH and NSPHT. The severity of alteration in the biochemical phenotype correlates with the type of mutation. Null mutations that prevent CaR expression cause mild FHH when heterozygous, but cause NSPHT when homozygous or compound heterozygous. Heterozygous mutations that permit CaR expression but impair function might cause more severe FHH or NSPHT by acting as dominant negatives of the wild-type CaR, presumably through heterodimerization. Truncation of the hCaR proximal to residue 888 disrupts receptor function; thus, frameshift and nonsense mutations causing such truncation are inactivating mutations. Missense mutations causing FHH/NSPHT might inactivate the CaR

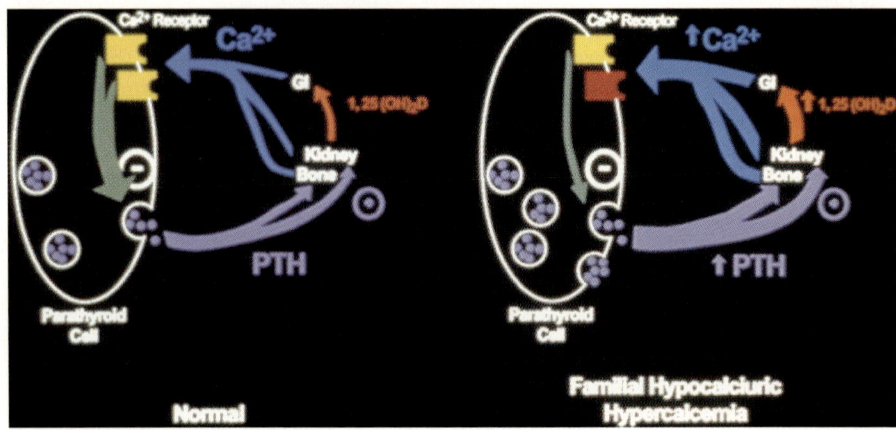

Fig. 7. Schematic depiction of a parathyroid cell with plasma membrane-localized CaR (*yellow*) and PTH-containing intracellular secretory granules (*white outline containing purple dots*). Extracellular Ca²⁺ activates the CaR, thereby inhibiting PTH secretion (*thick green arrow*). When extracellular Ca²⁺ decreases, PTH is secreted and acts directly on bone and kidney and indirectly on the gut (GI) through 1,25-dihydroxy vitamin D to raise serum Ca²⁺ in a normal homeostatic feedback loop. In familial hypocalciuric hypercalcemia, a loss-of-function mutation in one copy of the CaR gene (*red*) decreases the sensitivity of inhibition of parathyroid cell PTH secretion by Ca²⁺ (*thinner green arrow*) such that serum Ca²⁺ is increased, as is PTH secretion

by impairing normal folding and cell-surface expression or by preventing Ca²⁺ activation of the properly expressed receptor. Over 30 inactivating missense mutations in FHH/NSPHT have been identified to date, and their distribution is nonrandom. More than half cluster between residues 13 and 297 of the ECD, whereas only one has been reported between residues 298 and 548.

Heterozygous, activating mutations in subjects with ADH generally cause a left-shift in the Ca²⁺ set point, leading to relative hypocalcemia (Fig. 8) and hypercalciuria. With the exception of an in-frame deletion, ser895-val1075, activating mutations in ADH are missense mutations. Such mutations presumably act by relieving inhibitory constraints that maintain the CaR in its inactive conformation. Most ADH mutations increase Ca²⁺ sensitivity rather than causing constitutive activation. As with naturally occurring inactivating mutations, ADH mutations are clustered in particular regions of the CaR (Fig. 6). Most occur at the presumptive dimer interfaces of lobe 1 (particularly those within loop 2 shaded in Fig. 6) and of lobe 2 (pro221leu, glu228gln and gln245arg). We have suggested that these mutations enhance Ca²⁺ sensitivity by facilitating agonist-induced dimer rotation.

Within the 7TM domain, a cluster of mutations at the junction of TM helices 6 and 7 suggests that movement of these helices relative to each other could be a crucial event in CaR activation (Hu et al. 2005). Artificial mutation of proline 823 in TM6 (a residue highly conserved in family 3 GPCRs) to alanine drastically impairs Ca²⁺ activation of the receptor, despite intact expression of the mutant receptor at the cell surface. In contrast, a unique ADH mutation, ala843glu in TM7, leads to constitutive activation of the CaR, even when expressed in the ECD-deleted Rho-C-hCaR. These mutations further underscore the key role of TM6 and 7 in CaR activation.

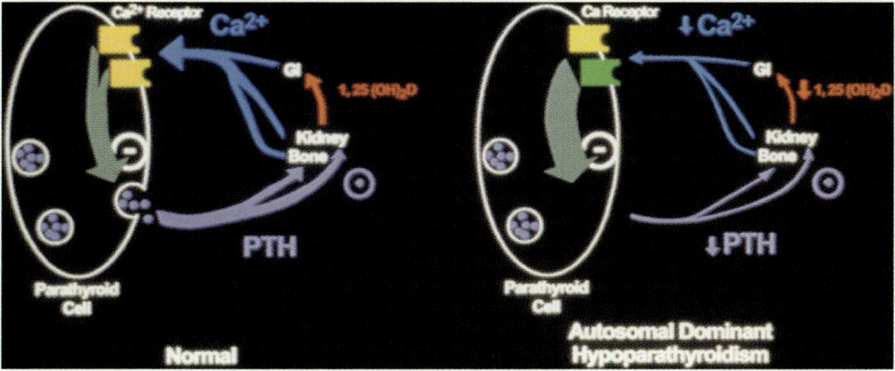

Fig. 8. Schematic depiction of a parathyroid cell as in Fig. 5. In autosomal dominant hypoparathy-roidism/hypocalcemia, a gain-of-function mutation in one copy of the CaR gene (*green*) increases the sensitivity of inhibition of parathyroid cell PTH secretion by Ca^{2+} (*thicker green arrow*) such that serum Ca^{2+} is decreased, as is PTH secretion

Allosteric modulators of the CaR

The central role of the CaR in regulating PTH secretion has made it an attractive target for positive and negative allosteric modulators, so-called calcimimetic and calcilytic drugs, respectively. Positive allosteric modulators of the CaR inhibit PTH secretion and could be useful in treatment of secondary hyperparathyroidism (e.g., in end-stage renal disease), in parathyroid cancer and in other forms of primary hyperparathyroidism not amenable to surgical treatment (Nemeth and Fox 1999). Negative allosteric modulators would increase PTH secretion and with appropriate pharmacokinetics could be useful as anabolic agents for treatment of osteoporosis (Nemeth et al. 2001).

Phenylalkylamines such as NPS R-568 act as positive allosteric modulators of the CaR, enhancing its sensitivity to Ca^{2+} without activating it by themselves. They are selective for the CaR, failing to modulate closely related family 3 GPCRs such as mGluR1. Presumably, selectivity reflects sequence differences at the drug-binding site, which has been shown to be within the 7TM domain. In particular, glutamate 837 has been identified as critical for binding of both positive and negative allosteric modulators such as NPS 568 and NPS 2143 (Hu et al. 2005). Since both of these compounds share a positively charged central amine, direct interaction with the negatively charged side chain of glutamate 837 may be critical for drug binding. Similarities between the action of the negative allosteric modulator, NPS 2143, and the pro823ala mutation in TM6 suggest that negative modulators may constrain the 7TM domain in a conformation that "resists" activation by signal transmitted from the agonist-bound VFT (Hu et al. 2005).

In in vitro studies, NPS 2143 inhibited Ca^{2+} activation of mutant forms of the CaR corresponding to those identified in subjects with ADH. Since patients with ADH are often hypercalciuric and at risk for development of kidney stones when treated with vi-tamin D and calcium to correct hypocalcemia, negative allosteric modulators might be particularly useful in treatment of ADH. Further studies are needed to test the possibil-ity that treatment of such patients with negative allosteric modulators would increase serum PTH and Ca^{2+} without the hypercalciuria seen with conventional treatment.

Conclusions and future studies

Studies of naturally occurring mutations of the hCaR have provided substantial insight into the structure and function of this unique GPCR. Its pivotal role in maintenance of extracellular Ca^{2+} homeostasis has spurred development of positive and negative allosteric modulators, some of which have already proved useful clinically. Further study of hCaR mutations and of novel allosteric modulators, combined with efforts to model the structure of the CaR 7TM domain and modulator binding site(s), should prove fruitful in helping us understand the mechanism of CaR activation and in developing more potent and selective drugs to modulate CaR activity.

Acknowledgements. I am grateful to many fellows in my own laboratory and collaborators from other laboratories who have contributed to our studies of the CaR. I would especially like to thank Dr. Jianxin Hu, staff scientist in my laboratory, Dr. Ken Jacobson and his colleagues in the Laboratory of Bioorganic Chemistry, NIDDK, who synthesized many of the compounds we studied, and Dr. Stefano Mora of Milan who identified many of the activating mutations of the CaR we studied.

References

Brown EM, MacLeod RJ (2001) Extracellular calcium sensing and extracellular calcium signaling. Physiol Rev 81:239–297

Brown EM, Gamba G, Riccardi D, Lombardi M, Butters R, Kifor O, Sun A, Hediger MA, Lytton J, Hebert SC (1993) Cloning and characterization of an extracellular Ca^{2+}-sensing receptor from bovine parathyroid. Nature 366:575–580

Bockaert J, Pin JP (1999) Molecular tinkering of G protein-coupled receptors: an evolutionary success. EMBO J 18:1723–1729

Chattopadhyay N, Yamaguchi T, Brown EM (1998) Ca^{2+} receptor from brain to gut: common stimulus, diverse actions. Trends Endocrinol Metab 9:354–359

Hu J, Spiegel AM (2003) Naturally occurring mutations of the extracellular Ca++-sensing receptor: implications for understanding its structure and function. Trends Endocrinol Metab 14:282–8

Hu J, McLarnon SJ, Mora S, Jiang J, Thomas C, Jacobson KA, Spiegel AM (2005) A region in the seven-transmembrane domain of the human Ca++ receptor Critical for response to Ca++. J Biol Chem 280:5113–5120

Kunishima N, Shimada Y, Tsuji Y, Sato T, Yamamoto M, Kumasaka T, Nakanishi S, Jingami H, Morikawa K (2000) Structural basis of glutamate recognition by a dimeric metabotropic glutamate receptor. Nature 407:971–977

Nemeth EF, Delmar EG, Heaton WL, Miller MA, Lambert LD, Conklin RL, Gowen M,Gleason JG, Bhatnagar PK, Fox J (1999) Calcimimetic compounds: a direct approach to controlling plasma levels of parathyroid hormone in primary hyperparathyroidism. Trends Endocrinol Metab 10:66–71

Nemeth EF, Delmar EG, Heaton WL, Miller MA, Lambert LD, Conklin RL, Gowen M, Gleason JG, Bhatnagar PK, Fox J (2001) Calcilytic compounds: potent and selective Ca^{2+} receptor antagonists that stimulate secretion of parathyroid hormone. J Pharm Exp Ther 299:323–331

Palczewski K, Kumasaka T, Hori T, Behnke CA, Motoshima H, Fox BA, Le Trong I, Teller DC, Okada T, Stenkamp RE, Yamamoto M, Miyano M (2000) Crystal structure of rhodopsin: a G protein-coupled receptor. Science 289:739–745

Spiegel AM, Weinstein LS (2004) Inherited disorders of G proteins and G protein-coupled receptors. Ann Rev Med 55:27–39

A molecular dissection
of the glycoprotein hormone receptors

Gilbert Vassart[1,2], *Leonardo Pardo*[3], and *Sabine Costagliola*[1]

Summary

In glycoprotein hormone receptors, a subfamily of rhodopsin-like G protein-coupled receptors, the recognition and activation steps are carried out by separate domains of the proteins. Specificity of recognition of the hormones [thyrotropin (TSH), lutropin (LH), human chorionic gonadotropin (hCG), follitropin (FSH)] involves leucine-rich repeats (LRR) present in an aminoterminal ectodomain and can be associated with a limited number of residues at key positions of the LRRs. The mechanism by which binding of the hormones results in activation is proposed to involve switching of the ectodomain from a tethered inverse agonist to a full agonist of the serpentine, rhodopsin-like portion of the receptor. Unexpectedly, the picture is complicated by the observation that promiscuous activation of one of the receptors (FSHr) by hCG or TSH can result from activating mutations affecting the serpentine portion of the receptors.

Introduction

The glycoprotein hormones and their receptors constitute an interesting example of co-evolution. The hormones, follitropin (FSH), lutropin (LH), human chorionic go-nadotropin (hCG) and thyrotropin (TSH), are dimeric proteins of about 30 kDa made of a common alpha subunit and specific beta subunits. The beta subunits are encoded by paralogous genes displaying substantial sequence similarity (Fig. 1a). The corresponding receptors, FSHr, LH/CGr and TSHr, are members of the rhodopsin-like, G protein-coupled receptor family. As such, they contain a "serpentine" portion containing seven transmembrane helices with many (but not all) of the sequence signatures typical of this receptor family. In addition, displaying a hallmark of the subfamily of glycoprotein hormone receptors (GPHRs), they contain a large (350–400 residues) aminoterminal ectodomain responsible for the high affinity and selective binding of the corresponding hormones (Fig. 2; Ascoli et al. 2002; Dias and Van Roey 2001; Szkudlinski et al. 2002). The higher sequence identity of the serpentine portions (about 70%) when compared with the ectodomains (about 40%, Fig. 1b) suggested early on that the former are interchangeable modules capable of activating the G proteins (mainly $G_{\alpha s}$) after specific binding of the individual hormones to the latter (Braun et al. 1991; Nagayama et al. 1990). Contrary to other rhodopsin-like GPCRs, binding of the hormones to their ectodmains can be observed with high affinity in the absence of the serpentine

[1] IRIBHM, Université Libre de Bruxelles, B-1070 Brussels, Belgium
[2] Service de Génétique Médicale, Hôpital Erasme, B-1070 Brussels, Belgium
[3] Laboratori de Medicina Computacional, Unitat de Bioestadística, Facultat de Medicina, Universitat Autònoma de Barcelona, 08193 Bellaterra, Spain

Conn et al.
Insights into Receptor Function
and New Drug Development Targets
© Springer-Verlag Berlin Heidelberg 2006

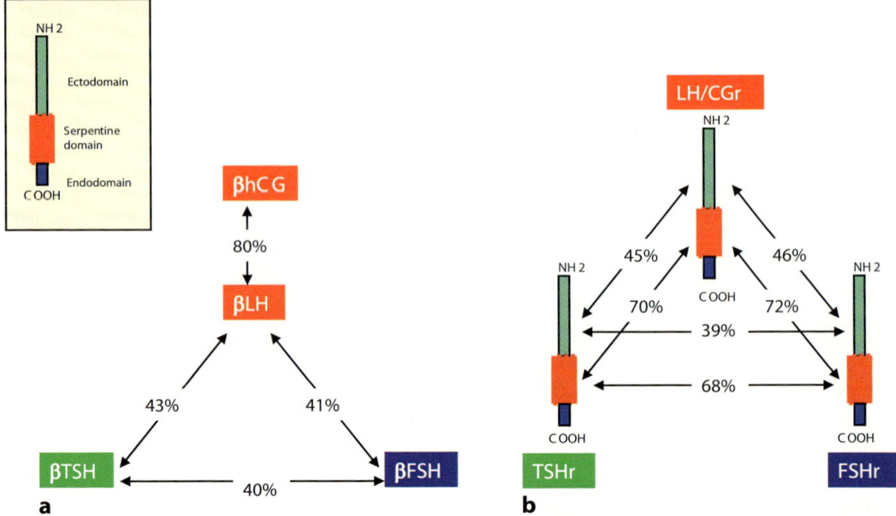

Fig. 1. The beta subunits of both glycoprotein hormones (**a**) and glycoprotein hormone receptors (**b**) are encoded by paralogous genes. Sequence identities are indicated separately for the ectodomains and serpentine domains of the three receptors and for the beta subunits of the four hormones. The pattern of shared similarities suggests co-evolution of the hormones and the ectodomain of their receptors, resulting in generation of specificity barriers. The high similarity displayed by the serpentine portions of the receptors is compatible with a conserved mechanism of intramolecular signal transduction

(Cornelis et al. 2001; Remy et al. 2001; Schmidt et al. 2001). The intramolecular transduction of the signal between these two portions of the receptors raises an interesting mechanistic issue which will be addressed in this essay.

Another issue, of evolutionary nature, relates to the shaping of hormone-receptor couples to cope with the emergence of chorionic gonadotropin in primates (Szkudlinski et al. 2002). Whereas in all mammals the circulating concentrations of TSH, LH and FSH are in proportion with their Kds for the corresponding receptors (in the low nanomolar range), in primates and, in particular, in man, chorionic gonadoropin (hCG), which shares its receptor with LH, can approach micromolar concentrations during the first trimester of pregnancy. This situation constitutes a challenge to the specificity of recognition regarding the TSH and FSH receptors and is known to be responsible for some spill-over in gestational trophoblastic disease, where even higher plasma levels of hCG are observed than in normal pregnancy.

The recognition step

Three-dimensional structures are available for hCG and FSH (Fox et al. 2001; Lapthorn et al. 1994; Wu et al. 1994), whereas for the ectodomains of the receptors, we are left with structural models covering only part of their structure (Fig. 2a,b). These are based on the known 3D structure of proteins containing leucine-rich repeats (LRRs; Bhowmick

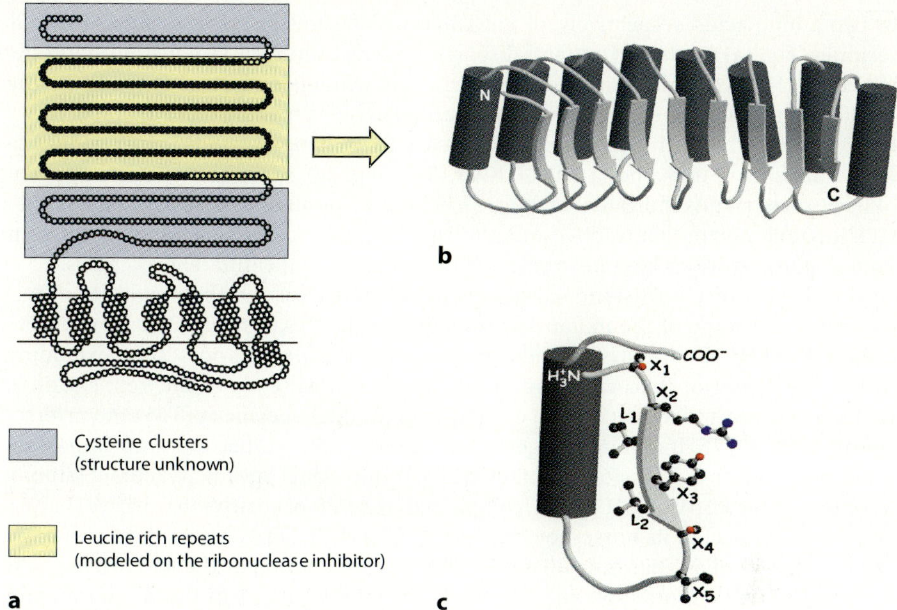

Fig. 2. Schematic representation of the structures composing glycoprotein hormone receptors. (a) Two-dimensional representation with indication of the various domains. The *blue boxes* correspond to aminoterminal and carboxyl terminal cysteine-rich portions of the ectodomain, flanking nine leucine-rich repeats (LRR, *yellow box*). (b) Repeats are made of 20 to 24 amino acids forming a β-strand followed by an α-helix. In LRR-containing proteins, the repeat units are arranged with their β-strands and α-helices parallel to a common axis and are organized spatially to form a horseshoe-shaped molecule, with the β-strands and α-helices making the concave and convex surfaces of the horseshoe, respectively. (c) Representation of a single LRR unit. The inner surface of the horseshoe is composed of seven residues: $X_1X_2LX_3LX_4X_5$. The side chains of the leucine residues are pointing inside the hydrophobic core of the protein and are important for its stability. The side chains of the X residues are predicted to be exposed to the solvent, making the surface available for interaction with the ligand (Kobe and Kajava 2001)

et al. 1996; Jiang et al. 1995; Kajava et al. 1995). LRRs are 20- to 25-residue protein motifs made of a beta strand and an alpha helix, connected by a turn (Fig. 2c). When assembled sequentially in a protein, the LRRs determine a "horseshoe-like" structure, with the beta strands making a concave inner surface (Fig. 2b). This surface has been shown to constitute the binding interface in the first LRR protein that has been crystallized: the ribonuclease inhibitor (Kobe and Deisenhofer 1993). Nine such motifs are found in the ectodomains of GPHRs. The model structure of the LRR portions of receptors predicts that non-leucine residues ($X_{1,2,3,4,5}$, Fig. 2c) would be pointing outwards and be available for interaction with the hormones, immediately suggesting that they might be implicated in recognition specificity. Flanking the LRR portion of the ectodomain are two cysteine-rich domains, the 3D structure of which is completely unknown.

Extensive amino acid substitutions by site-directed mutagenesis of the X_i residues in the LRR portion of the TSHr and the FSHr for their counterparts in the LH/CGr provided strong support to the above model (Smits et al. 2003a). Exchanging eight or two amino acids, respectively, of the TSHr or FSHr for the corresponding LH/CGr residues (Fig. 3a) resulted in mutants displaying a sensitivity to hCG matching that of the wild type LH/CGr (Smits et al. 2003a). Surprisingly, while gaining sensitivity to hCG, the mutants kept a normal sensitivity to their natural agonist, making them dual specificity receptors. For the TSH receptor, it is necessary to exchange 12 additional residues to fully transform it into a bona fide LH/CGr (T90, Fig. 3c). From an evolutionary point of view, these results indicate that nature has built recognition specificity of hormone-GPHR couples on both attractive and repulsive residues, and that residues at different homologous positions have been selected to this result in the different receptors.

Inspection of electrostatic surface maps of models of the three wild type (wt) receptors and some of the mutants is revealing in this respect (Fig. 3b,c; Smits et al. 2003a). The LH/CGr displays an acidic groove in the middle of its horseshoe, extending to the lower part of it (corresponding to the C-terminal ends of the beta strands). Generation of a similar distribution of charges in the dual-specificity (T56) and reverse-specificity (T90) TSHr mutants suggests that pattern of charge distribution is important for hCG recognition. Transformation of the FSHr into a dual-specificity mutant (simply by mutating the K104 and K179) is accompanied by a shift of positive to negative charges at the bottom part of its horseshoe (Fig. 3b,c). This shift suggests that these two basic residues function as guards against promiscuous recognition of the wt FSHr by hCG (Smits et al. 2003a). Attempts to correlate charge distributions in the wt and mutant receptors with those of residues in the three hormones suggest that the "seatbelt" portion of the beta subunits (known to play a key role in recognition specificity; Dias et al. 1994; Grossman et al. 1997; Moyle et al. 1994) might face the bottom border of the horseshoe.

As indicated above, the extremely high plasmatic concentrations of hCG, during the first trimester of pregnancy constitute a challenge to the recognition specificity of GPHRs. This has been illustrated recently in two pathological situations resulting from spontaneous mutations in the TSH and FSH receptors. In the first, mutation K183R, in the fifth LRR of the TSHr ectodmain, caused illegitimate stimulation of the thyroid gland by hCG, which resulted in severe hyperthyroidism during pregnancy (Rodien et al. 1998). In the framework of the discussion above, the consequences of this highly conservative amino acid substitution on the electrostatic surface map of the mutant receptor provided a structural rationale to the observed loss of specificity (Smits et al. 2002). In the second pathological situation, two mutant FSH receptors were identified, displaying promiscuous activation by hCG, in patients with spontaneous ovarian hyperstimulation syndrome (Smits et al. 2003b; Vasseur et al. 2003). Surprisingly, in these cases, the mutant residues were located in the serpentine portion of the receptor, outside the hormone binding domain, indicating that functional specificity might be distinct from true recognition specificity (see below).

In addition to the hormone-specific interactions genetically encoded in the primary structure of the ligands and the LRR portion of the receptors, we have recently demonstrated the importance of non hormone-specific ionic interactions involving sulphated tyrosines present in the ectodomains of all three receptors (Fig. 4a; Costagliola et al. 2002b). Similar to the interaction between von Willebrand factor and Gp1bα (Dong et al. 2001), sulfation of the three GPHRs contributes crucially to the binding affinity for the hormones without interfering with specificity. Whether the sulfated tyrosine residues interact directly with the hormones or contribute to correct shaping

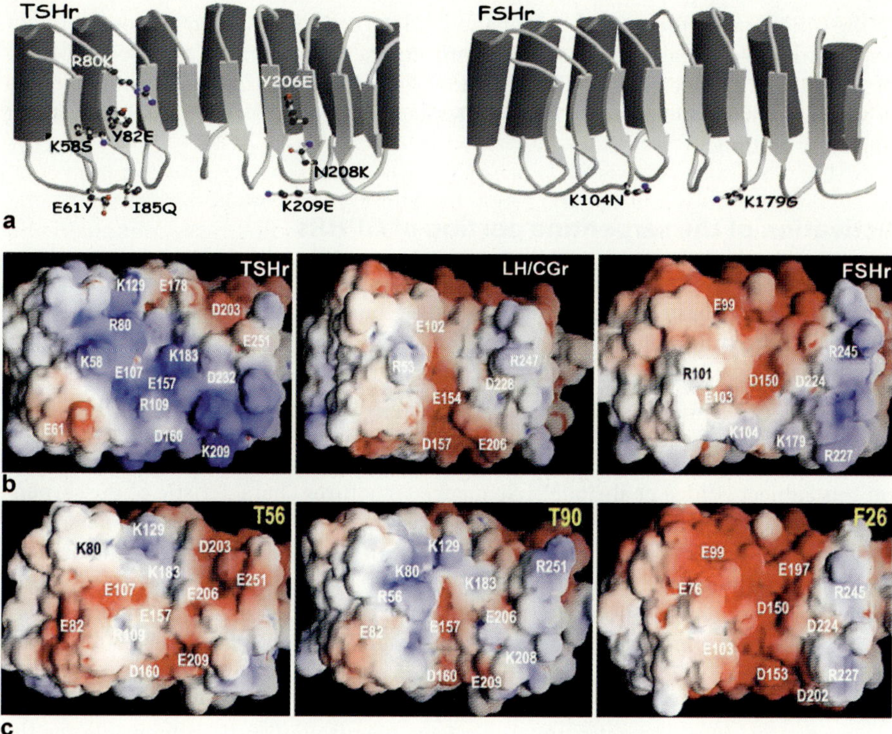

Fig. 3. Exchange of recognition specificity of glycoprotein hormone receptors by substitution of selected residues in the beta strands of the leucine-rich repeats (LRR) in their ectodomains (adapted from Smits et al. 2003a). (**a**) Schematic representation of the LRR portions of the thyrotropin receptor (TSHr, *left*) and follitropin receptor (FSHr, *right*), with indications of the eight (T56 mutant, *left*) or two residues (F26 mutant, *right*) leading to increases in sensitivity to recombinant human chorionic gonadotropin (rhCG). (**b**) Molecular electrostatic potential at the accessible surface of the models for the three wild type receptors. Note the acidic grove (represented in *red*) characteristic of the lower part of the middle portion in the ectdomain of the LH/CGr. (**c**) Molecular electrostatic potential at the accessible surface of the models established for TSHr and FSHr substitution mutants. Note the important reduction of electropositivity of the surface of T56, especially at the lower part of the middle portion of the horseshoe. This mutant displays sensitivity to rhCG similar to the LH/CGr, while keeping nominal sensitivity to recombinant humanTSH (Smits et al. 2003a). T90 mutant is a TSHr with 20 residues of the beta strands of the LRR portion exchanged with the LH/CGr. Its molecular electrostatic potential at the accessible surface reproduces very closely the acidic groove of the wild type LH/CGr. This mutant is as sensitive to rhCG as the LH/CGr and has completely lost sensitivity to recombinant human TSH (Smits et al. 2003a). The molecular electrostatic potential at the surface of the F26 mutant displays also an acidic groove. The main difference with the wild type FSHr is the extension of the electronegative region to the lower middle portion of the horseshoe. The F26 mutant behaves functionally as a dual receptor, with increased sensitivity to rhCG and conserved sensitivity to rhFSH (Smits et al. 2003a)

of the functional ectodomains remains to be determined. A similar reasoning might apply to the role of sialylated or sulfated carbohydrate chains linked to the ectodomain of GPHRs. In this respect, it is worth noting that amino acid substitutions in the ectodomain of GPHRs could, in theory, alter carbohydrate structure and, hence, affect hormone binding.

Activation of the serpentine portion of GPHRs

As they belong to the rhodopsin-like GPCR family and display many of the cognate signatures in primary structure, the serpentine portions offix-start-drag GPHRs are likely to share common mechanisms of activation with rhodopsin. Crystallographic data are only available for the inactive conformation of rhodopsin (Palczewski et al. 2000). Nevertheless, molecular scenarios for the activation phenomenon have been proposed, based on a panel of experimental approaches involving site-directed muta-genesis, cross-linking and molecular modeling. [Readers are referred to a recent review on this subject (Ridge et al. 2003).] Sequence signatures characteristic of the serpentine portion of GPHR suggest, however, the existence of idiosyncrasies associated with their specific mechanisms of activation. In addition, over the past 10 years, the LH/CGr and, even more so, the TSHr have been found to be activated by a wide spectrum of gain of function mutations (Parma et al. 1997; Refetoff et al. 2001; Shenker 2002). In LH/CGr, such mutations cause pseudoprecocious puberty (Shenker 2002), a rare disease expressed only in males and transmitted on the autosomal dominant mode. In TSHr, in addition to germline mutations causing hereditary toxic thyroid hyperplasia (Duprez et al. 1994), somatic mutations have been found to be responsible for the majority of autonomous thyroid adenomas (Parma et al. 1997), a relatively frequent and easily diagnosed condition. Compilation of data available for both receptors has identified more than 30 residues, the mutation of which causes constitutive activation. As many somatic mutations affecting a given residue have been found repeatedly in the TSHr (and do not involve hypermutable targets), it is likely that we are approaching a saturation map for spontaneous gain of function mutations. Attempts to translate this map into mechanisms of transition between inactive and active conformations of

Fig. 4. Sequence signatures common to all rhodopsin-like G protein-coupled receptors and sequence signatures specific to the glycoprotein hormone receptor (GPHR) gene family are both implicated in activation of GPHRs. (**a**) Linear representation of a typical GPHR, with indications of key residues (*red dots*) and conserved motifs. SO_3^{2-} stands for postranslational sulfation of the indicated tyrosine residues. The *black boxes* stand for transmembrane helices and I1-I3, E1-E3, for intracellular and extracellular loops, respectively. (**b**) Molecular modeling of the GPHR-specific lock between transmembrane helices VI and VII, involving mainly residues 6.43, 6.44 and 7.49. Rupture of this lock by mutation of D6.44 or D6.43 would release N7.49, making its side chain available for productive interactions (possibly with D2.50) and resulting in constitutive activation (Govaerts et al. 2001). 1a, 1b and 1c stand for water molecules, the position of which is modeled from the structure of rhodopsin (Okada et al. 2002). (**c**) Molecular modeling of the ionic lock predicted to exist between the D/ERY/W motif at the bottom of transmembrane helix III and an acidic residue in position 6.30 (D567 in the follitropin receptor). Rupture of this interaction also causes constitutive activation (Parma et al. 1993; Smits et al. 2003b)

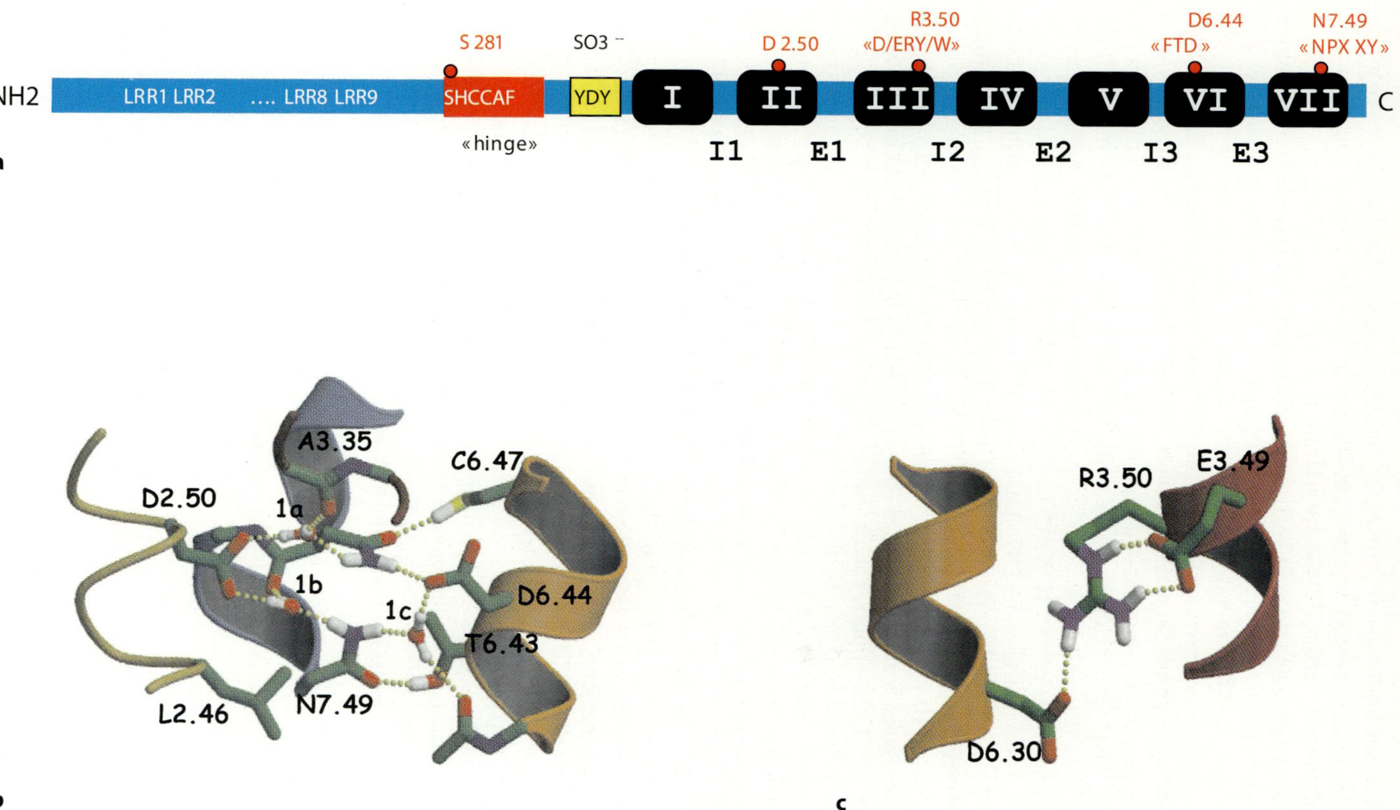
a
NH2
LRR1 LRR2 .... LRR8 LRR9
SHCCAF
«hinge»
YDY
S 281
SO3 −
I
II
III
IV
V
VI
VII
C
I1 E1 I2 E2 I3 E3
D 2.50
R3.50
«D/ERY/W»
D6.44
«FTD»
N7.49
«NPX XY»
b
D2.50
A3.35
1a
1b
1c
C6.47
D6.44
T6.43
N7.49
L2.46
c
R3.50
E3.49
D6.30

the receptors are underway, in light of the rhodopsin structural data. Three sequence patterns affected by gain of function mutations deserve special mention and might help us to understand how GPHRs are activated.

First, aspartate 6.44 (D633 in TSHr; D578 in LH/CGr) belongs to the "FTD signature", a motif specific to GPHRs, at the cytoplasmic side of transmembrane helix VI (TM-VI; Fig. 4a). When mutated to a variety of amino acids, including alanine, the result is constitutive activation in both the TSHr and LH/CGr (Govaerts et al. 2001; Lin et al. 1997; Neumann et al. 2001; Parma et al. 1997). This finding suggested early on that the gain of function resulted from the breakage of (a) bond(s), rather than the creation of novel interaction(s) by the mutated residue. The observation, in a GPHR homologue of Drosophila, of a reciprocal mutation involving D6.44 of the "FTD" motif in TM-VI and N7.49 of the "NPXXY" motif in TM-VII (Fig. 4a) suggested that an interaction between D6.44 and N7.49 would exist in the inactive conformation of GPHRs (Govaerts et al. 2001). Figure 4b shows a molecular model in which a set of water molecules observed in the D2.50/N7.49 environment of Rhodopsin has been included (indicated as 1a, 1b and 1c in Fig. 4b; Okada et al. 2002). Constitutive activation would thus be the consequence of breakage of the interaction between T6.43 or D6.44 and N7.49 (Fig. 4b). Interestingly, TSHr constructs bearing the N7.49A substitution can no longer be activated by TSH, despite normal expression and binding of the hormone (Govaerts et al. 2001). We tentatively conclude that, in the inactive conformation of GPHRs, the side chain of N7.49 is normally "sequestered" by both T6.43 and D6.44, and that the active conformation(s) require(s) establishment of novel interaction(s) of N7.49. Asparagine 7.49 of the NPXXY motif is one of the most conserved residues in rhodopsin-like GPCRs. It has been suggested, on the basis of a reciprocal substitution with D2.50 in the GnRH receptor, that N7.49 might be implicated in the activation mechanism (Sealfon et al. 1995; Zhou et al. 1994) via creation of an interaction with D2.50. The experimental data on rhodopsin structure have recently confirmed this suggestion (Fritze et al. 2003). Our observations are compatible with this hypothesis. They suggest that, in GPHRs, evolution has selected a novel motif in TM-VI to control an activation switch common to all rhodopsin-like receptors. Whether this is related to the peculiarities of the activation mechanism of GPHRs, involving their large ectodomain (see below), remains to be determined.

Second, glutamate 3.49 and arginine 3.50 of the highly conserved "D/ERY/W" motif at the bottom of TM-III form an ionic lock with aspartate 6.30 at the cytoplasmic end of TM-VI (Fig. 4a,c). Disruption of this ionic lock by either the E3.49A or E3.49Q mutations or mutations affecting D6.30 leads to constitutively active mutant receptors (Claeysen et al. 2002). Thus, the movement of TM-III and TM-VI at the cytoplasmic side of the membrane is necessary for receptor activation (Ballesteros et al. 2001).

Third, serine 281 belongs to a GPHR-specific "YPSHCCAF" sequence signature located downstream of the LRR portion, in the ectodomain of the receptors (Fig. 4a). After mutation of this serine residue had been shown to activate the TSH receptor constitutively (Duprez et al. 1997; Kopp et al. 1997), this segment, sometimes referred to as the "hinge" motif, was shown to play an important role in activation of all three GPHRs (Nakabayashi et al. 2000). The functional effect of substitutions of S281 in the TSHr, or S277 in the LH/CGr, likely results in a "loss-of-structure", locally, since the more de-structuring the substitutions, the stronger the activation (Ho et al. 2001; Nakabayashi et al. 2000). This observation, together with results showing that mutation of specific residues in the extracellular loops of the TSHr causes constitutive activation (Parma et al. 1995), led to the hypothesis that activation of the receptor could result from the rupture of an inhibitory interaction between the ectodomain and the serpentine domain (Duprez et al. 1997).

Finally, it is important to note that N7.49 of the "NPxxY" motif seems to be involved in stabilization of both the inactive and active states of the receptor, since the basal activity levels of all activating mutations, including those affecting the D/ERY/W motif at the cytoplasmic side or S281 in the ectodomain, were significantly decreased by suppression of the side-chain of N7.49 (N7.49A double mutants; Claeysen et al. 2002).

Interaction between the ectodomain and the serpentine domain

The hypothesis that the ectodomain would exert an inhibitory effect on an inherently noisy rhodopsin-like serpentine domain was supported by early data showing that mild treatment of TSHr-expressing cells by trypsin caused partial activation of the receptor (Van Sande et al. 1996). Definite demonstration of such an effect was made by Zhang et al. (2000), who showed activation of the TSHr secondary to "beheading", in aminoterminal truncated mutants. However, careful comparison of the activity of truncated mutants with maximally stimulated wt TSHr, or S281 gain of function mutants indicated that truncation of the receptor of its ectodomain resulted only in partial activation of the serpentine domain (Vlaeminck et al. 2002). In addition, engineering activating mutations in the serpentine of an aminoterminally truncated mutant resulted in further activation of the constructs (Vlaeminck et al. 2002). Interestingly, only mutations in the transmembrane helices were effective; substitutions in the extracellular loops of serpentine-only constructs were without effects (Vlaeminck et al. 2002).

From these observations, we proposed the following model for activation of the TSHr (Fig. 5a; Vlaeminck et al. 2002). In the resting state, the ectodomain would exert an inhibitory effect on the activity of an inherently noisy rhodopsin-like serpentine, qualifying pharmacologically as an inverse agonist of the serpentine. Upon activation, by binding of the hormone, or secondary to mutation of S281 in the hinge region, the ectodomain would switch from inverse agonist to full agonist of the serpentine portion. The ability of the strongest S281 mutants to fully activate the receptor in the absence of hormone suggests that the ultimate agonist of the serpentine domain would be the "activated" ectodomain, with no need for an interaction between the hormone and the serpentine domain. The ineffectiveness of mutations in the extracellular loops to activate serpentine-only constructs suggests that, in the wt receptor, the exoloops and a portion of the ectodomain (the hinge region?) do cooperate in the generation of a structural module functioning as an agonist of the serpentine. Results obtained with chimeras between the FSHr and LGRs agree with this model and suggest that the second exoloop plays a key role in this mechanism (Nishi et al. 2002). It is noteworthy that direct interaction of the hormone with the serpentine is not required in this model, albeit it is by no means excluded.

Unexpected complications

According to an appealing evolutionary scenario, and fitting with the functional data analyzed above, GPHRs would have evolved from two distinct genes: the first, encoding a typical rhodopsin-like serpentine receptor, the second encoding multiple LRR domains involved in protein-protein interactions (Gross et al. 1991). The ancestral GPHR gene must be extremely old, as a similar genomic organization is found in LGRs, members of which are already present in sea anemones (Nothacker and Grimmelikhuijzen 1993). All this points to a neat dichotomy between hormone recognition (by the LRRs in the ectodomain) and activation of the G protein (by the serpentine domain). This logic of division of labor fits well with the data reported above. It is also supported by the identification of natural and experimental mutations in the ectodomain of GPHRs, which affect specificity of hormone recognition (Rodien et al. 1998; Smits et al. 2003a). Recent data, however, challenge this view and make it necessary to revise the tight functional dichotomy between recognition and activation.

It is revealing that, here again, the way was paved by natural mutations identified in patients with an interesting phenotype. Two families were identified in which female patients presented with spontaneous ovarian hyper stimulation syndrome (spontaneous OHSS) and mutation in their FSHr (Smits et al. 2003b; Vasseur et al 2003). In the majority of cases, this condition is caused by excessive stimulation of the ovaries by exogenous gonadotropins administered in the context of in vitro fertilization procedures (iatrogenic OHSS; Delvigne and Rozenberg 2002). In these two families, the disease occurred spontaneously on the occasion of each pregnancy. In both cases, it was shown that the mutated FSH receptors were abnormally sensitive to the pregnancy hormone hCG, thus providing a satisfactory explanation for the phenotype. The surprise was that the amino acid substitutions were located in the serpentine portion of the FSHr, rather than affecting the LRR portion, as would be expected (Fig. 5, panel b (iii)).

Fig. 5. Interactions between the ectodomain and the serpentine domains are implicated in the activation mechanism and functional specificity. The receptors are represented with their ectodomain containing a concave, hormone-binding structure facing upwards, and a transmembrane serpentine portion. (**a**) Our current model for activation of the thyrotropin receptor (TSHr). The basal state of the receptor is characterized by an inhibitory interaction between the ectodomain and the serpentine domain (indicated by the ϡ (−) *blue sign*). The ectodomain would function as a tethered inverse agonist of the serpentine portion. Mutation of Ser281 in the ectodomain into leucine switches the ectodomain from an inverse agonist into a full agonist of the serpentine domain (indicated by the ϡ (+) *red sign*). Binding of TSH (indicated by αβ dimeric structure) to the ectodomain is proposed to have a similar effect, converting it into a full agonist of the serpentine portion (adapted from (Vlaeminck et al. 2002)). The serpentine portions in the basal (silent) state are represented as compact, *black structures*. Fully activated serpentine portions are depicted as relaxed *red structures*, with *arrows* indicating activation of Gαs. (**b**) Our current interpretation of the phenotypes displayed by the D567N and T449I follitropin receptor (FSHr) mutants of patients with spontaneous ovarian hyperstimulation syndrome (OHSS; Smits et al. 2003b; Vasseur et al. 2003), in light of the model in panel (**a**). High affinity interactions of the ectodomain of the wild type FSHr with FSH would result in full activation of the serpentine portion (**b i**). High concentrations of human chorionic gonadotropin (hCG), while capable of establishing low affinity interactions with the ectodomain of the FSHr, would be ineffective in activating a strongly locked wild type serpentine domain (**b ii**). In both D567N and T449I OHSS mutants (location of the mutations schematically indicated by *yellow dots*), partial activation of the serpentine domains (indicated as a partially relaxed structure, in *blue*) would lower the activation threshold of the serpentine, thus allowing for stimulation to take place (**b iii**)

One of the mutations, D567N (Smits et al. 2003b), affects residue 6.30 (see above and Fig. 4c). The phenotype of the mutant receptors was studied in more detail by additional site-directed mutagenesis and functional assays in COS cells, and the following picture emerged (Smits et al. 2003b): 1) in addition to allowing promiscuous stimulation by hCG, both mutations caused increases in the constitutive activity of a normally silent FSHr; 2) the loss of specificity was not restricted to hCG, with the mutants also showing

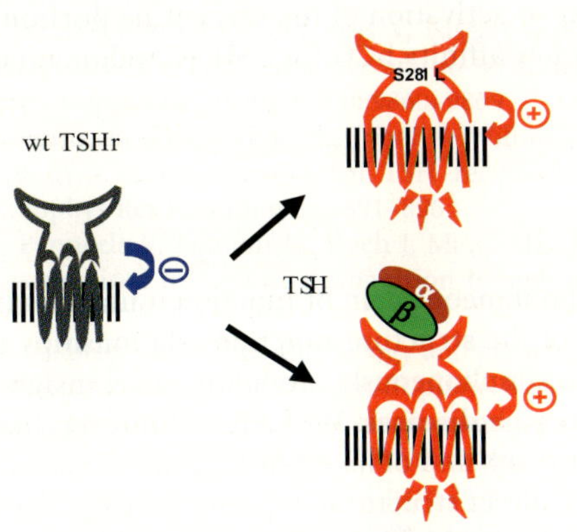

a

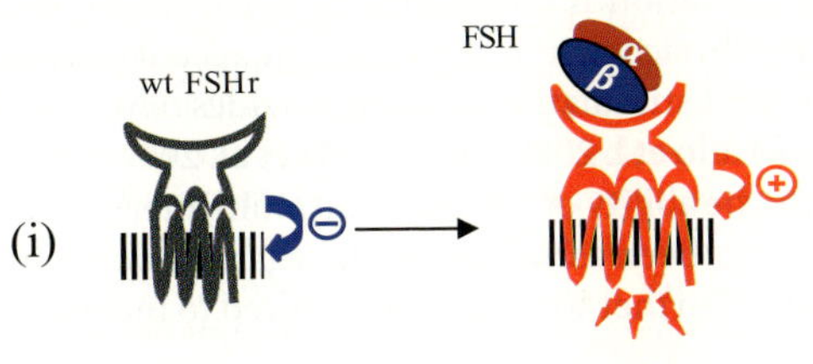

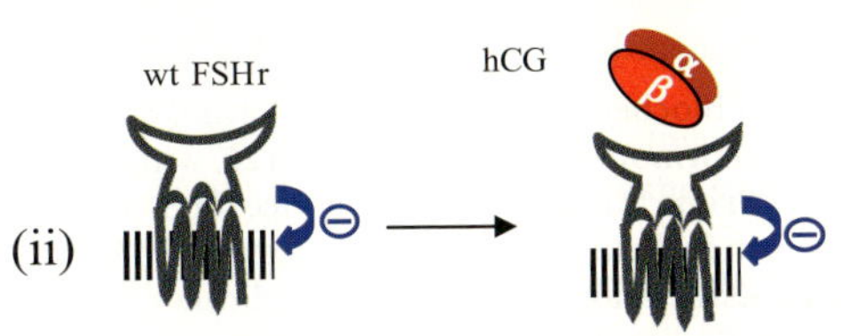

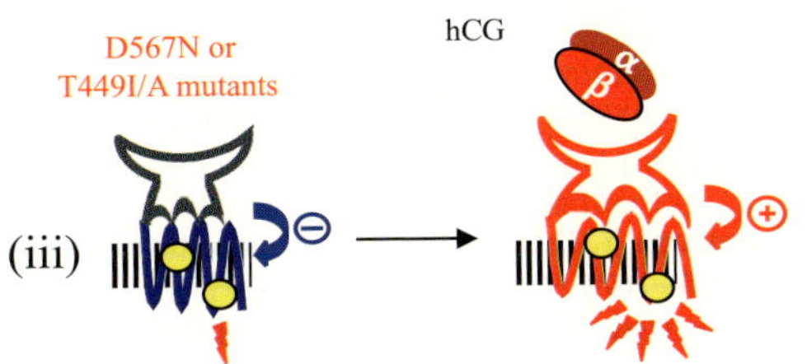

b

Ji I, Lee C, Song Y, Conn PM, Ji TH (2002) Cis- and trans-activation of hormone receptors: the LH receptor. Mol Endocrinol 16:1299–1308

Jiang X, Dreano M, Buckler DR, Cheng S, Ythier A, Wu H, Hendrickson WA, el Tayar N (1995) Structural predictions for the ligand-binding region of glycoprotein hormone receptors and the nature of hormone-receptor interactions. Structure 3:1341–1353

Kajava AV, Vassart G, Wodak SJ (1995) Modeling of the three-dimensional structure of proteins with the typical leucine-rich repeats. Structure 3:867–877

Kobe B, Deisenhofer J (1993) Crystal structure of porcine ribonuclease inhibitor, a protein with leucine-rich repeats. Nature 366:751–756

Kobe B, Kajava AV (2001) The leucine-rich repeat as a protein recognition motif. Curr Opin Structn Biol 11:725–732

Kopp P, Muirhead S, Jourdain N, Gu WX, Jameson JL, Rodd C (1997) Congenital hyperthyroidism caused by a solitary toxic adenoma harboring a novel somatic mutation (serine281->isoleucine) in the extracellular domain of the thyrotropin receptor. J Clin Invest 100:1634–1639

Kudo M, Osuga Y, Kobilka BK, Hsueh AJ (1996) Transmembrane regions V and VI of the human luteinizing hormone receptor are required for constitutive activation by a mutation in the third intracellular loop. J Biol Chem 271:22470–22478

Lapthorn AJ, Harris DC, Littlejohn A, Lustbader JW, Canfield RE, Machin KJ, Morgan FJ, Isaacs NW (1994) Crystal structure of human chorionic gonadotropin, Nature 369:455–461

Lee C, Ji I, Ryu K, Song Y, Conn PM, Ji TH (2002) Two defective heterozygous luteinizing hormone receptors can rescue hormone action. J Biol Chem 277:15795–15800

Lin Z, Shenker A, Pearlstein R (1997) A model of the lutropin/choriogonadotropin receptor: insights into the structural and functional effects of constitutively activating mutations. Protein Eng 10:501–510

Moyle WR, Campbell RK, Myers RV, Bernard MP, Han Y, Wang X (1994) Co-evolution of ligand-receptor pairs. Nature 368:251–255

Nagayama Y, Russo D, Chazenbalk GD, Wadsworth HL, Rapoport B (1990) Extracellular domain chimeras of the TSH and LH/CG receptors reveal the mid-region (amino acids 171–260) to play a vital role in high affinity TSH binding, Biochem Biophys Res Commun 173:1150–1156

Nakabayashi K, Kudo M, Kobilka B, Hsueh AJ (2000) Activation of the luteinizing hormone receptor following substitution of Ser-277 with selective hydrophobic residues in the ectodomain hinge region. J Biol Chem 275:30264–30271

Neumann S, Krause G, Chey S, Paschke R (2001) A free carboxylate oxygen in the side chain of position 674 in transmembrane domain 7 is necessary for TSH receptor activation. Mol Endocrinol 15:1294–1305

Nishi S, Nakabayashi K, Kobilka B, Hsueh AJ (2002) The ectodomain of the luteinizing hormone receptor interacts with exoloop 2 to constrain the transmembrane region: studies using chimeric human and fly receptors. J Biol Chem 277:3958–3964

Nothacker HP, Grimmelikhuijzen CJ (1993) Molecular cloning of a novel, putative G protein-coupled receptor from sea anemones structurally related to members of the FSH, TSH, LH/CG receptor family from mammals. Biochem Biophys Res Commun 197:1062–1069

Okada T, Fujiyoshi Y, Silow M, Navarro J, Landau EM, Shichida Y (2002) Functional role of internal water molecules in rhodopsin revealed by X-ray crystallography. Proc Natl Acad Sci USA 99:5982–5987

Palczewski K, Kumasaka T, Hori T, Behnke CA, Motoshima H, Fox BA, Le T, I, Teller DC, Okada T, Stenkamp RE, Yamamoto M, Miyano M (2000) Crystal structure of rhodopsin: A G protein-coupled receptor. Science 289:739–745

Parma J, Duprez L, Van Sande J, Cochaux P, Gervy C, Mockel J, Dumont JE, Vassart G (1993) Somatic mutations in the thyrotropin receptor gene cause hyperfunctioning thyroid adenomas. Nature 365:649–651

Parma J, Van Sande J, Swillens S, Tonacchera M, Dumont JE, Vassart G (1995) Somatic mutations causing constitutive activity of the TSH receptor are the major cause of hyperfunctional thyroid adenomas: identification of additional mutations activating both the cAMP and inisitolphosphate-Ca++ cascades. Mol Endocrinol 9:725–733

Parma J, Duprez L, Van Sande J, Hermans J, Van Vliet G, Costagliola S, Rodien P, Dumont JE, Vassart G (1997) Diversity and prevalence of somatic mutations in the TSH receptor and Gs alpha genes as a cause of toxic thyroid adenomas. J Clin Endocrinol Metab 82:2695–2701

Refetoff S, Dumont JE, Vassart G (2001) Thyroid disorders. In: Scriver CR, Beaudet AL, Sly WS, Valle eds. The metabolic and molecular bases of inherited diseases. McGraw-Hill, New York, pp 4029–4076

Remy JJ, Nespoulous C, Grosclaude J, Grebert D, Couture L, Pajot E, Salesse R (2001) Purification and structural analysis of a soluble human chorionogonadotropin hormone-receptor complex. J Biol Chem 276:1681–1687

Ridge KD, Abdulaev NG, Sousa M, Palczewski K (2003) Phototransduction: crystal clear. Trends Biochem Sci 28:479–487

Rodien P, Bremont C, Sanson ML, Parma J, Van Sande J, Costagliola S, Luton JP, Vassart G, Duprez L (1998) Familial gestational hyperthyroidism caused by a mutant thyrotropin receptor hypersensitive to human chorionic gonadotropin. New Engl J Med 339:1823–1826

Sanders J, Jeffreys J, Depraetere H, Richards T, Evans M, Kiddie A, Brereton K, Groenen M, Oda Y, Rees Smith B (2002) Thyroid stimulating monoclonal antibodies. Thyroid. 12:1043–50

Schmidt A, MacColl R, Lindau-Shepard B, Buckler DR, Dias JA (2001) Hormone-induced conformational change of the purified soluble hormone binding domain of follitropin receptor complexed with single chain follitropin. J Biol Chem 276:23373–23381

Sealfon SC, Chi L, Ebersole BJ, Rodic V, Zhang D, Ballesteros JA, Weinstein H (1995) Related contribution of specific helix 2 and 7 residues to conformational activation of the serotonin 5-HT2A receptor. J Biol Chem 270:16683–16688

Shenker A (2002) Activating mutations of the lutropin choriogonadotropin receptor in precocious puberty. Receptors Channels 8:3–18

Smits G, Govaerts C, Nubourgh I, Pardo L, Vassart G, Costagliola S (2002) Lysine 183 and glutamic acid 157 of the thyrotropin receptor: two interacting residues with a key role in determining specificity towards TSH and hCG. Mol Endocrinol 16:722–35

Smits G, Campillo M, Govaerts C, Janssens V, Richter C, Vassart G, Pardo L, Costagliola S (2003a) Glycoprotein hormone receptors: determinants in leucine-rich repeats responsible for ligand specificity. EMBO J 22:2692–2703

Smits G, Olatunbosun O, Delbaere A, Pierson R, Vassart G, Costagliola S (2003b) Ovarian hyperstimulation syndrome due to a mutation in the follicle-stimulating hormone receptor. New Engl J Med 349:760–766

Szkudlinski MW, Fremont V, Ronin C, Weintraub BD (2002) Thyroid-stimulating hormone and thyroid-stimulating hormone receptor structure-function relationships. Physiol Rev 82:473–502

Urizar E, Montanelli L, Loy T, Bonomi M, Swillens S, Gales C, Bouvier M, Smits G, Vassart G, Costagliola S (2005) Glycoprotein hormone receptors: link between receptor homodimerization and negative cooperativity 1, EMBO J 24:1954–1964

Van Sande J, Massart C, Costagliola S, Alleier A, Cetani F, Vassart G, Dumont JE (1996) Specific activation of the thyrotropin receptor by trypsin. Mol Cell Endocrinol 119:161–168

Vasseur C, Rodien P, Beau I, Desroches A, Gerard C, de Poncheville L, Chaplot S, Savagner F, Croue A, Mathieu E, Lahlou N, Descamps P, Misrahi M (2003) A chorionic gonadotropin-sensitive mutation in the follicle-stimulating hormone receptor as a cause of familial gestational spontaneous ovarian hyperstimulation syndrome. New Engl J Med 349:753–759

Vlaeminck V, Ho SC, Rodien P, Vassart G, Costagliola S (2002) Activation of the cAMP pathway by the TSH receptor involves switching of the ectodomain from a tethered inverse agonist to an agonist. Mol Endocrinol 16:736–746

Wu H, Lustbader JW, Liu Y, Canfield RE, Hendrickson WA (1994) Structure of human chorionic gonadotropin at 2.6 A resolution from MAD analysis of the selenomethionyl protein. Structure 2:545–558

Zhang M, Tong KP, Fremont V, Chen J, Narayan P, Puett D, Weintraub BD, Szkudlinski MW (2000) The extracellular domain suppresses constitutive activity of the transmembrane domain of the human TSH receptor: implications for hormone-receptor interaction and antagonist design. Endocrinology 141:3514–3517

Zhou W, Flanagan C, Ballesteros JA, Konvicka K, Davidson JS, Weinstein H, Millar RP, Sealfon SC (1994) A reciprocal mutation supports helix 2 and helix 7 proximity in the gonadotropin-releasing hormone receptor. Mol Pharmacol 45:165–170

Receptor Tyrosine Kinases as Targets for Cancer Therapy Development

Andreas Gschwind[1], *Oliver M. Fischer*[1], and *Axel Ullrich*[1]

Summary

Receptor tyrosine kinases (RTKs) are membrane-spanning proteins that possess a ligand-controlled intracellular kinase activity. They regulate a wide variety of cellular processess as diverse as cell proliferation, apoptosis or cell migration. Consequently, dysregulation of RTKs due to overexpression, mutation or autocrine stimulation has been causally linked to cancer development and progression. The advent of molecular cloning allowed the elucidation of the primary structure of the first RTK, the EGFR. Subsequent research in this field led to tremendous advances in understanding molecular signalling processes governing both physiological and pathophysiological behaviour of cells. These discoveries paved the way for the development of target-specific cancer therapeutics and opened up a new era of molecular targeted approaches in the treatment of human cancer. The approval of monoclonal antibodies such as Herceptin© for the treatment of breast cancer or small molecule inhibitors such as Gleevec© for gastrointestinal stromal tumors underlines both the power and success of this novel strategy.

Introduction

In the 20 years since the first isolation and characterization of the cDNA of the EGFR (EGF receptor), intensive research efforts have led to important insights into the molecular mechanisms of receptor tyrosine kinase (RTK) action. Moreover, substantial advances have been made in understanding the key roles of RTKs in the signaling pathways that govern fundamental cellular processes, such as cell proliferation, migration, metabolism, differentiation and survival, and that regulate intercellular communication during development. RTK activity in resting, normal cells is tightly controlled. When mutated or structurally altered, however, RTKs become potent oncoproteins. Abnormal activation of RTKs in transformed cells has been shown to be causally involved in the development and progression of human cancers. Consequently, RTKs and their growth factor ligands have become rational targets for therapeutic intervention by humanized antibodies and small molecule drugs. In recent years, RTK-based cancer therapies for the treatment of, for example, metastatic breast cancer, gastrointestinal stromal tumors and non-small cell lung cancer have reached widespread clinical use and have thereby demonstrated the power of gene-based therapy development.

[1] Max-Planck-Institute of Biochemistry Department of Molecular Biology, Am Klopferspitz 18 82152 Martinsried, Germany

Conn et al.
Insights into Receptor Function
and New Drug Development Targets
© Springer-Verlag Berlin Heidelberg 2006

The Advent of Molecular Cloning in Growth Factor Signaling Research

The development of gene technology in the mid-1970s led to a major breakthrough in the field of growth factor signalling research. The technology had already allowed the identification of the cDNAs that encoded important physiological peptide hormones and growth factors, such as insulin (Ullrich et al. 1977; Sures et al. 1980), EGF (Gray et al. 1983; Scott et al. 1983), insulin-like growth factor (IGF)-2 (Dull et al. 1984), NGF (Ullrich et al. 1983), PDGF (Johnsson et al. 1984; Chiu et al. 1984) and transforming growth factor alpha (TGFα; Derynck et al. 1984). This identification, in turn, led to their sequencing and the determination of their amino acid sequences. Incidentally, the ability to produce peptides like somatostatin (Itakura et al. 1977) in bacteria and later even manufacture medically important hormones, including insulin and growth hormone (Martial et al. 1979), on a large scale spawned the biotech industry in the late 1970s.

As cDNA cloning technologies improved during the early 1980s, it became feasible to clone large gene transcripts, so several laboratories directed their efforts towards the elucidation of a cell surface receptor that mediates the mitogenic activity of the growth factor EGF. It was widely expected that this accomplishment would significantly improve the understanding of the mechanisms that regulate basic biological phenomena, such as cell proliferation and differentiation of both normal and transformed cells. In 1984, a team of collaborators from the Imperial Cancer Research Fund (ICRF), Genentech and The Weizmann Institute of Science isolated and characterized the cDNA sequence of the human EGFR – the prototypical RTK – from normal placental and A431 tumor cells (Ullrich et al. 1984).

Insights from the EGFR Primary Structure

The first peptide sequences of purified EGFR immediately caused a sensation. Julian Downward – who at that time was in Michael Waterfield's laboratory at the ICRF – searched known protein sequences for matches and hit the jackpot (Downward et al. 1984). He found a very high level of similarity between the EGFR peptides and sequences of an avian oncogene, v-erbB, which had been reported shortly before by Tadashi Yamamoto (Yamamoto et al. 1983). This discovery connected for the first time an animal oncogene with a human gene that encoded a cell growth-controlling membrane protein, that is, a gene product with partially known normal functions. Further detailed information was obtained (Ullrich et al. 1984) from cloning and sequencing the complete EGFR cDNA. Truncations, deletions and mutations in the v-erbB oncogene were identified that, as it was speculated then, were found to be the genetic bases of the conversion of a proto-oncogene into an oncogene that can cause malignant cancer in avian erythroblastosis virus-infected chicken. Moreover, the characterization of the EGFR cDNA revealed the first complete amino acid sequence of a cell surface receptor with signal-generating capacity and provided detailed insights into its molecular architecture. The human EGFR was found to be a large glycoprotein with a modular structure: it included an extracellular ligand binding domain, a transmembrane region and an intracellular cytoplasmic tyrosine kinase portion that is flanked by noncatalytic regulatory regions. The EGFR cDNA cloning project (Ullrich et al. 1984) yielded two other important discoveries. First, Southern blot analysis with an EGFR cDNA probe revealed a 25-fold amplification of the EGFR gene in human A431 epidermal carcinoma cells, a prototypical genetic abnormality that should prove to be of key relevance for future developments. Second, the screening of cDNA libraries yielded EGFR-related, but clearly distinct, cDNA sequences that gained importance in subsequent studies. The EGFR gene family is now known to comprise four members: the EGFR, human EGFR-related (HER) 2, kinase-impaired HER3 and HER4.

In the years following the characterization of the EGFR primary structure, the nucleotide sequences and deduced primary amino acid sequences of several other RTKs were reported by the Genentech lab and its collaborators, including the InsR (Ullrich et al. 1985; Ebina et al. 1985), the insulin-like growth factor I receptor (IGF-IR; Ullrich et al. 1986), the PDGFR (Yarden et al. 1986), as well as of the proto-oncogenes c-kit (Yarden et al. 1987) and c-fms (Coussens et al. 1986). This series of studies verified that, in spite of their unique biological roles, RTKs are highly related in structure and share a domain arrangement that is very similar to that of the EGFR. The RTK class of cell surface receptors now comprises 58 members that are distributed into 20 subfamilies, and more than half of the known RTKs have been found to be overexpressed or mutated in human hyper- or hypo-proliferative diseases (Blume-Jensen and Hunter 2001).

The Role of the EGFR Family in Cancer

Starting with the discovery of the protooncogene-oncogene connection between the EGFR and v-erB, intense efforts were undertaken to investigate a potential role of the EGFR in human cancer. In the 1980s, numerous reports described the overexpression of the EGFR in a variety of epithelial tumors and substantiated the view that dysregulated EGFR signalling plays an important role in human cancers. Following these observations, enhanced stimulation of the EGFR through autocrine growth factor loops, in particular via TGFα (Sizeland and Burgess 1992), was identified as a common mechanism of RTK deregulation. Moreover, many laboratories embarked on a massive search for EGFR mutations in human cancers, and several deletions and point mutations were described that result in enhanced catalytic tyrosine kinase activity of the receptor (Humphrey et al. 1990). The most prevalent of these mutations in tumors was found to be EGFRvIII, an EGFR deletion mutant lacking exons 2–7 that can arise from gene rearrangement or alternative mRNA splicing (Malden et al. 1988).

In 1985, the complete primary structure of a putative RTK that displayed high homology to the human EGFR, named human EGFR-related (HER) 2 (Coussens et al. 1985), was reported. Other laboratories also discovered this new EGFR relative with unknown function and they designated it c-erbB2 (King et al. 1985). Interestingly, the chromosomal localization of HER2 was identical to that of the rat neu oncogene (Schechter et al. 1984), which provided another connection between an RTK and cancer development in animals. The oncogenic significance of neu was further substantiated when Robert Weinberg and co-workers showed that monoclonal antibodies against the neu oncogene reverted its transforming effects in NIH3T3 cells (Drebin et al. 1984). The critical next step, which addressed the key question as to whether genetic abnormalities in the EGFR or HER2 system could be identified in human tumors, was taken through a collaboration formed in 1985 by the Ullrich lab and Dennis Slamon, an oncologist at

UCLA. Slamon had assembled a collection of primary breast tumors and was ready to use Ullrich's gene probes to search for abnormalities in tumor DNA. Two years later, the collaborators reported that the HER2 gene is amplified in 30% of invasive breast cancers and, for the first time, revealed a significant correlation between HER2 overexpression in tumors and reduced patient survival and time to relapse (Slamon et al. 1987). These findings established HER2 as a prognostic factor and suggested a critical role of HER2 overexpression in the pathogenesis of breast and ovarian cancer (Slamon et al. 1989).

RTKs as Targets in Cancer Therapy: The Development of Herceptin…

The discovery of HER2 gene amplification in breast and ovarian cancer provided an opportunity to evaluate the concept of target-specific cancer therapy. The Genentech group set out to develop HER2-specific monoclonal antibodies (MAb) and to assess their anti-oncogenic potential in cell culture and animal model systems (Hudziak et al. 1989; Fendly et al. 1990). This work provided the basis for the subsequent humanization of MAb 4D5 and the development of the therapeutic antibody trastuzumab (Herceptin®, Genentech Inc.) as the first genomic research-based, targeted anti-kinase therapeutic agent (Table 1). Herceptin was approved by the US Food and Drug Administration (FDA) for the treatment of HER2-overexpressing metastatic breast cancer in 1998 (Table 2). It binds HER2 on the surface of tumor cells (Hudziak et al. 1989) and induces receptor internalization, inhibition of cell cycle progression and recruitment of immune effector cells. Demonstration of the anti-tumor activity of Herceptin in breast cancer patients provided the proof of principle that therapy targeted against a human oncoprotein in a major cancer indication could be successful.

First described by Hudziak and colleagues (Hudziak et al. 1989) in 1989, the anti-HER2 antibody 2C4, named pertuzumab in its humanized form (Omnitarg™, rhuMAb-2C4, Genentech Inc.), that is currently in phase II clinical development represents a second generation of anti-HER2 monoclonal antibody that interferes with the mechanism of oncogenic signal generation by HER2–HER3 heterodimers (Agus et al. 2002) and, thereby, complements the molecular armamentarium to fight cancers with low levels of HER2 expression.

… and Erbitux

A visionary effort and landmark accomplishment in EGFR-targeted cancer therapy was the design of the mouse monoclonal antibodies 225 and 528 to extracellular epitopes of the receptor (Kawamoto et al. 1983; Sato et al. 1983) by Mendelsohn and colleagues in the early 1980s. Based on its promising anti-tumor activity in cultured human tumor cell lines and rodent models, the 225 antibody was selected for clinical development. And, in 2003, the Swiss Agency for Therapeutic Products (Swissmedic) approved the use of the chimeric human–mouse anti-EGFR antibody Cetuximab (IMC-C225®, Erbitux, ImClone Systems/Merck KGaA) for the treatment of patients with colorectal cancer who no longer respond to standard chemotherapy treatment with irinotecan. The fact that it took 20 years to develop C225 as a therapeutic exemplifies the many pitfalls that can significantly affect the realization of a novel concept in clinical application.

Table 1. RTK-targeted cancer therapies

Name	Target	Indication	Description	Company
Trastuzumab, Herceptin	HER2/neu	Breast cancer	Humanized, anti-HER2 IgG1 kappa	Genentech
Imatinib, Glivec, STI1571	Bcr-Abl, c-kit, PDGFR	CML GIST	2-Phenylaminopyrimidine	Novartis
Gefitinib, Iressa, ZD1839	EGFR	NSCLC	Quinazoline	AstraZeneca
Cetuximab, Erbitux	EGFR	Colorectal cancer	Chimeric anti-EGFR IgG1	ImClone/ Merck
Pertuzumab, Omnitarg, 2C4	HER2	Clinical development	Humanized anti-HER2 (heterodimerization inhibitor)	Genentech
SU6668	VEGFR2, PDGFR, FGFR	Clinical development	Indoline-2-one	SUGEN/ Pfizer
SU11248	VEGFR2, c-kit, PDGFR, FLT3	Glivec-resistant GIST	Indoline-2-one	SUGEN/ Pfizer
ZD6474	VEGFR2	Clinical development	Quinazoline	AstraZeneca
PTK-787	VEGFR1 +2	Clinical development	Anilinophthalazine	Novartis/ Schering
Befacizumab, Avastin	VEGF	Clinical development	Humanized anti-VEGF (rhuMAb-VEGF)	Genentech
Nexavar	Raf VEGFR+	RCC		Bayer

Abbreviations: Abl, Abelson tyrosine kinase; Bcr, breakpoint cluster region protein; CML, chronic myelogenous leukemia; EGFR, epidermal growth factor receptor; FGFR, fibroblast growth factor receptor; FLT, fms-related tyrosine kinase; GIST, gastrointestinal stromal tumors; HER, human EGFR related; Ig, immunoglobulin, NSCLC, non-small-cell lung carcinoma; PDGFR, platelet-derived growth factor receptor; RTK, receptor tyrosine kinase; VEGFR, vascular endothelial growth factor receptor.

Small Molecule Inhibitors of RTKs

After EGFR and other tyrosine kinases had been validated as suitable pharmacological targets for anti-cancer drugs, one of the hottest races in pharmaceutical development began: to identify rationally designed, small molecule, anti-cancer drugs. Levitzki's group at the Hebrew University in Jerusalem was at the forefront in the development of tyrosine kinase inhibitors that were targeted to RTKs and demonstrated their potential use as antiproliferative agents (Margolis et al. 1989; Yaish et al. 1988) in the late 1980s. The therapeutic approach to target the EGFR with small-molecule inhibitors is based on the early observations by Honegger and co-workers in 1987 that mutations in the ATP-binding pocket of the EGFR abrogates the tyrosine kinase func-

Table 2. Timeline: breakthrough discoveries on RTK signal transduction and RTK-based cancer therapy

1952	Discovery of NGF
1962	Discovery of EGF
1977	Cloning of rat insulin cDNA as a first step towards medical application of gene technology
1978	Identification of the EGFR
1979	Phosphorylation of proteins on tyrosine is discovered
1980	EGFR is recognized as a protein tyrosine kinase
1983	Cloning of pre-pro-EGF cDNA
1983	The EGFR is first targeted with the mouse monoclonal antibodies 225 and 528 to block proliferation of human cancer cell lines
1984	Cloning of the EGFR reveals homology to the v-erb-B oncogene; discovery of EGFR gene amplification in human cervix carcinoma cell line
1984	EGF stimulates GTP binding to RAS
1985	Characterization of the EGFR-related HER2 (c-erbB2) gene
1985	Cloning of cDNA encoding the InsR
1986	Characterization of a chimeric EGFR–InsR receptor indicates that RTKs use similar mechanisms to transmit signals across the plasma membrane
1987	The HER2 gene is found to be amplified in 30% of invasive breast cancers; significant clinical correlation between HER2 over-expression and poor clinical outcome
1989	Phospholipase Cg is discovered to be a substrate of PDGFR and EGFR; identification of SH2 and SH3 domains
1990	PI3K is discovered to be associated with the stimulated EGFR
1992	cDNA cloning of VEGFR1 (FLT1) and VEGFR2 (FLK1)
1993	Knockout of TGFa is found to result in abnormal skin architecture, wavy hair, curly whiskers and eye abnormalities
1993	Anti-VEGF antibodies are demonstrated to inhibit the growth of cancer cells in nude mice
1994	Retroviral gene transfer of a dominant negative FLK1 mutant blocks the in vivo growth of glioblastoma tumors in nude mice
1995	Knockout of the EGFR is found to result in severe defects in epithelial development of multiple organs
1996	GPCRs are demonstrated to transactivate the EGFR and HER2 signal
1998	Trastuzumab (Herceptin®) is approved by the FDA for the treatment of HER2-overexpressing breast cancer in the US
2000	Trastuzumab (Herceptin®) is approved in European countries
2001	Imatinib (Glivec®) receives approval for use in patients with CML in the US
2002	Imatinib (Glivec®) receives approval for use in patients with advanced GIST in the US
2002	Gefitinib (Iressa®) is approved in Japan for the treatment of inoperable and recurrent NSCLC
2002	High resolution crystal structures of the extracellular portion of the EGFR with EGF or TGFa establishes a receptor-mediated mechanism of receptor dimerization
2003	Gefitinib (Iressa®) is approved in the US for the treatment of advanced NSCLC

Table 2. (continued)

2003	Cetuximab (Erbitux™) is approved in Switzerland for the treatment of patients with advanced colorectal cancer
2005	Nexavar approval by FDA for RCC
2006	SUTENT approved for treatment of Glivec-resistant GIST

tion of the receptor (Honegger et al. 1987) and that these mutations interfere with EGFR oncogenic signalling (Honegger et al. 1987; Redemann et al. 1992). In 1994, the tyrosine kinase inhibitory activities of quinazolines were first described (Fry et al. 1994; Osherov and Levitzki 1994), and two years later, Wakeling and co-workers reported the pharmacological characteristics of gefitinib (Iressa® ZD1839, AstraZeneca) as a potent and selective inhibitor of the EGFR tyrosine kinase activity(Wakeling et al. 1996). In 2002, gefitinib was approved in Japan for the treatment of inoperable and recurrent non-small-cell lung carcinoma (NSCLC); it was approved one year later in the US.

Several pharmaceutical companies and academic laboratories have successfully developed small-molecule tyrosine kinase inhibitors. Imatinib (Glivec®, STI571, Novartis), which was originally developed by Ciba scientists led by Alex Matter as a derivative of a protein kinase C inhibitor, has provided the "proof of concept" for the clinical efficacy and tolerability of this compound class and was the first selective inhibitor to be approved by the FDA for the treatment of cancer. Imatinib was first described in 1996 by Druker and colleagues (1996) as having potent activity against the Bcr-Abl. The constitutively activated non-receptor tyrosine kinase Bcr-Abl is found in chronic myelogenous leukemia (CML) cells that express the Philadelphia chromosome, a reciprocal translocation between chromosomes 9 and 22 that replaces the first exon of c-*abl* with sequences from the *bcr* gene. This translocation represents the key oncogenic event in 95% of patients with CML. In addition to Abl and Bcr-Abl, the RTKs PDGFR and c-kit were found to be potently inhibited by imatinib (Buchdunger et al. 2000). As c-kit is believed to have an important role in the pathogenesis of gastrointestinal stromal tumors (GIST), clinical studies with imatininb were successfully extended to this tumor type (Joensuu et al. 2001). Imatinib received approval by the FDA for use in patients with CML in 2001 and for advanced GIST in 2002.

Targeting Angiogenesis

Another strategy to inhibit the growth of cancer tumors involves the targeting of the signaling system that controls the formation of new blood vessels – angiogenesis. Vascular endothelial growth factor (VEGF) and its receptors are known to be important players in the regulation of tumor angiogenesis (Ferrara 2002; Folkman 1971). In 1992, DeFries discovered that FLT1 (FMS-like-tyrosine kinase) was a receptor for VEGF (De Vries et al. 1992) a second VEGF receptor, VEGFR2 (FLK1, KDR) was subsequently described (Terman et al. 1992; Millauer et al. 1993; Quinn et al. 1993). A crucial role for both of these RTKs in angiogenesis was demonstrated by knocking out the genes in mice (Fong et al. 1995; Shalaby et al. 1995). Proof that VEGF and

VEGFR signalling is required for tumor angiogenesis was presented in two seminal studies in the mid-1990s: Napoleone Ferrara and his associates showed that anti-VEGF antibodies abrogate the growth of tumor xenografts in nude mice (Kim et al. 1993), and Birgit Millauer and colleagues (1994) demonstrated that a dominant negative VEGFR2 mutant blocks the subcutaneous growth of experimental glioblastomas in the same model. Later, the broad relevance of this discovery was further substantiated by data obtained from a variety of other tumor types (Millauer et al. 1996). The use of retroviruses encoding dominant-interfering mutants of RTKs in this series of experiments suggested a therapeutic application of retroviral gene therapies in the treatment of human cancers. More importantly, however, the experimental results of Millauer and Ferrara demonstrated the clinical potential of anti-angiogenic therapy by targeting either the ligand or the corresponding receptor as critical elements of a biological signalling system.

Based on these findings, VEGF and VEGFRs became established as important targets for therapeutic intervention in tumor growth. VEGF was targeted by monoclonal neutralizing antibodies and VEGFR by small chemical compounds. Befacizumab (Avastin^{TM}, Genentech Inc.) is a humanized antibody against VEGF (Presta et al. 1997) that has been approved by the FDA for the treatment of colorectal and other cancers.

The first small-molecule VEGFR antagonist to enter clinical trials was SU5416 (SUGEN), which was later followed by SU6668. These compounds competitively block ATP binding to the tyrosine kinase domain of the receptor, thereby inhibiting tumor angiogenesis in vivo and the growth of xenografts established from a variety of human cancers (Fong et al. 1999; Shaheen et al. 1999). The related compound SU11248 targets multiple RTKs (O'Farrell et al. 2003), including c-kit, PDGFR, FLT3 and VEGFR2, and is currently being evaluated in clinical trials for the treatment of patients with a variety of cancers. The use of SU11248 (SUTENT) for the treatment of Glivec-resistant gastrointestinal stroma tumors as well as renal cell carcinoma was approved by the FDA in early 2006.

Conclusions

In the 20 years since the cloning of the first cDNA encoding an RTK, the EGFR, much progress has been made in our understanding of the fundamental signalling mechanisms of RTKs, their biology and the pathological consequences of RTK deregulation. Although a complete understanding of RTK function and dysfunction in diverse tissues and multiple biological processes is still to come, the work on members of this gene family has already had a major impact on cancer therapy. Trastuzumab, Imatinib, Gefitinib, Cetuximab, Nexavar and SUTENT have demonstrated the potential of targeted cancer therapeutics, and several other RTK-based, experimental anticancer strategies are now in late-stage clinical development. Important questions, such as the definition of the optimal dose and schedule of drug administration, as well as the issue of resistance formation that has been observed in Glivec-treated CML (Gorre et al. 2001) and GIST (Heinrich et al. 2003) patients, remain to be addressed. Industrial as well as academic research will further focus on evaluating RTKs as promising molecular targets for cancer treatment. An impressive example is a recent study in which high-throughput sequencing technologies, combined with bioinformatics, were used to systematically analyze the tyrosine kinome in colorectal cancer. The large-scale sequencing approach identified several previously unknown mutations in tyrosine kinase genes that could be targeted for therapeutic intervention in the future. Due to the extensive complexity of pathogenic alterations in the cancer cell signalling network, gene-based diagnostic techniques, such as gene array, tissue array and single-nucleotide polymorphism analysis, will help select patients who are likely to respond favorably to a particular anti-signalling drug. Ultimately, because of the plasticity of the genome of cancer cells, it will be essential to develop combination therapies involving small molecule and antibody cocktails that act through distinct and complementary mechanisms of action in order to achieve rapid and complete eradication of tumors.

References

Agus DB, Akita RW, Fox WD, Lewis GD, Higgins B, Pisacane PI, Lofgren JA, Tindell C, Maiese K, Scher HI, Sliwkowski MX (2002) Targeting ligand-activated ErbB2 signaling inhibits breast and prostate tumor growth. Cancer Cell 2:127–137

Bardelli A, Parsons DW, Silliman N, Ptak J, Szabo S, Saha S, Markowitz S, Willson JK, Parmigiani G, Kinzler KW, Vogelstein B, Velculescu VE (2003) Mutational analysis of the tyrosine kinome in colorectal cancers. Science 300:949

Blume-Jensen P, Hunter T (2001) Oncogenic kinase signalling. Nature 411: 355–365

Buchdunger E, Cioffi CL, Law N, Stover D, Ohno-Jones S, Druker BJ, Lydon NB (2000) Abl protein-tyrosine kinase inhibitor STI571 inhibits in vitro signal transduction mediated by c-kit and platelet-derived growth factor receptors. J Pharmacol Exp Ther 295:139–145

Chiu IM, Reddy EP, Givol D, Robbins KC, Tronick SR, Aaronson SA (1984) Nucleotide sequence analysis identifies the human c-sis proto-oncogene as a structural gene for platelet-derived growth factor. Cell 37:123–129

Coussens L, Yang-Feng TL, Liao YC, Chen E, Gray A, McGrath J, Seeburg PH, Libermann TA, Schlessinger J, Francke U, Ullrich A (1985) Tyrosine kinase receptor with extensive homology to EGF receptor shares chromosomal location with neu Oncogene. Science 230:1132–1139

Coussens L, Van Beveren C, Smith D, Chen E, Mitchell RL, Isacke CM, Verma IM, Ullrich A (1986) Structural alteration of viral homologue of receptor proto-oncogene fms at carboxyl terminus. Nature 320:277–280

Derynck R, Roberts AB, Winkler ME, Chen EY, Goeddel DV (1984) Human transforming growth factor-alpha:precursor structure and expression in E. coli. Cell 38:287–297

de Vries C, Escobedo JA, Ueno H, Houck K, Ferrara N, Williams LT (1992) The fms-like tyrosine kinase, a receptor for vascular endothelial growth factor. Science 255:989–991

Downward J, Yarden Y, Mayes E, Scrace G, Totty N, Stockwell P, Ullrich A, Schlessinger J, Waterfield MD (1984) Close similarity of epidermal growth factor receptor and v-erb-B oncogene protein sequences. Nature 307:521–527

Drebin JA, Stern DF, Link VC, Weinberg RA, Greene MI (1984) Monoclonal antibodies identify a cell-surface antigen associated with an activated cellular oncogene. Nature 312:545–548

Druker BJ, Tamura S, Buchdunger E, Ohno S, Segal GM, Fanning S, Zimmermann J, Lydon NB (1996) Effects of a selective inhibitor of the Abl tyrosine kinase on the growth of Bcr-Abl positive cells. Nat Med 2:561–566

Dull TJ, Gray A, Hayflick JS, Ullrich A (1984) Insulin-like growth factor II precursor gene organization in relation to insulin gene family. Nature 310: 777–781

Ebina Y, Ellis L, Jarnagin K, Edery M, Graf L, Clauser E, Ou JH, Masiarz F, Kan YW, Goldfine ID, Roth RA, Rutter WJ (1985) The human insulin receptor cDNA:the structural basis for hormone-activated transmembrane signalling. Cell 40:747–758

Fendly BM, Winget M, Hudziak RM, Lipari MT, Napier MA, Ullrich A (1990) Characterization of murine monoclonal antibodies reactive to either the human epidermal growth factor receptor or HER2/neu gene product. Cancer Res 50:1550–1558

Ferrara N (2002) VEGF and the quest for tumour angiogenesis factors. Nat Rev Cancer 2:795–803

Folkman J (1971) Tumor angiogenesis:therapeutic implications. New Engl J Med 285:1182–1186

Fong GH, Rossant J, Gertsenstein M, Breitman ML (1995) Role of the Flt-1 receptor tyrosine kinase in regulating the assembly of vascular endothelium. Nature 376:66–70

Fong TA, Shawver LK, Sun L, Tang C, App H, Powell TJ, Kim YH, Schreck R, Wang X, Risau W, Ullrich A, Hirth KP, McMahon G (1999) SU5416 is a potent and selective inhibitor of the vascular endothelial growth factor receptor (Flk-1/KDR) that inhibits tyrosine kinase catalysis, tumor vascularization, and growth of multiple tumor types. Cancer Res 59:99–106

Fry DW, Kraker AJ, McMichael A, Ambroso LA, Nelson JM, Leopold WR, Connors RW, Bridges AJ (1994) A specific inhibitor of the epidermal growth factor receptor tyrosine kinase. Science 265:1093–1095

Gorre ME, Mohammed M, Ellwood K, Hsu N, Paquette R, Rao PN, Sawyers CL (2001) Clinical resistance to STI-571 cancer therapy caused by BCR-ABL gene mutation or amplification. Science 293:876–880

Gray A, Dull TJ, Ullrich A (1983) Nucleotide sequence of epidermal growth factor cDNA predicts a 128,000-molecular weight protein precursor. Nature 303:722–725

Heinrich MC, Corless CL, Demetri GD Blanke CD, von Mehren M, Joensuu H, McGreevey LS, Chen CJ, Van den Abbeele AD, Druker BJ, Kiese B, Eisenberg B, Roberts PJ, Singer S, Fletcher CD, Silberman S, Dimitrijevic S, Fletcher JA (2003) Kinase mutations and imatinib response in patients with metastatic gastrointestinal stromal tumor. J Clin Oncol 21:4342–4349

Honegger AM, Dull TJ, Felder S, Van Obberghen E, Bellot F, Szapary D, Schmidt A, Ullrich A, Schlessinger J (1987) Point mutation at the ATP binding site of EGF receptor abolishes protein-tyrosine kinase activity and alters cellular routing. Cell 51:199–209

Honegger AM, Szapary D, Schmidt A, Lyall R, Van Obberghen E, Dull TJ, Ullrich A, Schlessinger J (1987) A mutant epidermal growth factor receptor with defective protein tyrosine kinase is unable to stimulate proto-oncogene expression and DNA synthesis. Mol Cell Biol 7 4568–4571

Hudziak RM, Lewis GD, Winget M, Fendly BM, Shepard HM, Ullrich A (1989) p185HER2 monoclonal antibody has antiproliferative effects in vitro and sensitizes human breast tumor cells to tumor necrosis factor. Mol Cell Biol 9:1165–1172

Humphrey PA, Wong AJ, Vogelstein B, Zalutsky MR, Fuler GN, Archer GE, Friedmann HS, Kwatra MM, Bigner SH, Bigner DD (1990) Anti-synthetic peptide antibody reacting at the fusion junction of deletion-mutant epidermal growth factor receptors in human glioblastoma. Proc Natl Acad Sci USA 87:4207–4211

Itakura K, Hirose T, Crea R, Riggs AD, Heyneker HL, Bolivar F, Boyer HW (1977) Expression in Escherichia coli of a chemically synthesized gene for the hormone somatostatin. Science 198:1056–1063

Joensuu H, Roberts PJ, Sarlomo-Rikala M, Andersson LC, Tervahartiala P, Tuveson D, Silberman S, Capdeville R, Dimitrijevic S, Druker B, Demetri GD (2001) Effect of the tyrosine kinase inhibitor STI571 in a patient with a metastatic gastrointestinal stromal tumor. N Engl J Med 344:1052–1056

Johnsson A, Heldin CH, Wasteson A, Westermark B, Deuel TF, Huang JS, Seeburg PH, Gray A, Ullrich A, Scrace G, Stroobant P, Waterfield MD (1984) The c-sis gene encodes a precursor of the B chain of platelet-derived growth factor. Embo J 3:921–928

Kawamoto T, Sato JD, Le A, Polikoff J, Sato GH, Mendelsohn J (1983) Growth stimulation of A431 cells by epidermal growth factor:identification of high-affinity receptors for epidermal growth factor by an anti-receptor monoclonal antibody. Proc Natl Acad Sci U S A 80:1337–1341

Kim KJ, Li B, Winer J, Armanini M, Gillett N, Phillips HS, Ferrara N (1993) Inhibition of vascular endothelial growth factor-induced angiogenesis suppresses tumour growth in vivo. Nature 362:841–844

King CR, Kraus MH, Aaronson SA (1985) Amplification of a novel v-erbB-related gene in a human mammary carcinoma. Science 229:974–6

Malden LT, Novak U, Kaye AH, Burgess AW (1988) Selective amplification of the cytoplasmic domain of the epidermal growth factor receptor gene in glioblastoma multiforme. Cancer Res 48:2711–2714

Margolis B, Rhee SG, Felder S, Mervic M, Lyall R, Levitzki A, Ullrich A, Zilberstein A, Schlessinger J (1989) EGF induces tyrosine phosphorylation of phospholipase C-II:a potential mechanism for EGF receptor signaling. Cell 57:1101–1107

Martial JA, Hallewell RA, Baxter JD, Goodman HM (1979) Human growth hormone: complementary DNA cloning and expression in bacteria. Science 205:602–607

Millauer B, Wizigmann-Voos S, Schnurch H, Martinez R, Moller NP, Risau W, Ullrich A (1993) High affinity VEGF binding and developmental expression suggest Flk-1 as a major regulator of vasculogenesis and angiogenesis. Cell 72:835–846

Millauer B, Shawver LK, Plate KH, Risau W, Ullrich A (1994) Glioblastoma growth inhibited in vivo by a dominant-negative Flk-1 mutant. Nature 367:576–579

Millauer B, Longhi MP, Plate KH, Shawver LK, Risau W, Ullrich A, Strawn LM (1996) Dominant-negative inhibition of Flk-1 suppresses the growth of many tumor types in vivo. Cancer Res 56:1615–1620

O'Farrell AM, Abrams TJ, Yuen HA, Ngai TJ, Louie SG, Yee KW, Wong LM, Hong W, Lee LB, Town A, Smolich BD, Manning WC, Murray LJ, Heinrich MC, Cherrington JM (2003) SU11248 is a novel FLT3 tyrosine kinase inhibitor with potent activity in vitro and in vivo. Blood 101:3597–3605

Osherov N, Levitzki A (1994) Epidermal-growth-factor-dependent activation of the src-family kinases. Eur J Biochem 225:1047–53

Presta LG, Chen H, O'Connor SJ, Chisholm V, Meng YG, Krummen L, Winkler M, Ferrara N (1997) Humanization of an anti-vascular endothelial growth factor monoclonal antibody for the therapy of solid tumors and other disorders. Cancer Res 57:4593–4599

Quinn TP, Peters KG, De Vries C, Ferrara N, Williams LT (1993) Fetal liver kinase 1 is a receptor for vascular endothelial growth factor and is selectively expressed in vascular endothelium. Proc Natl Acad Sci USA 90:7533–7537

Redemann N, Holzmann B, von Ruden T, Wagner EF, Schlessinger J, Ullrich A (1992) Anti-oncogenic activity of signalling-defective epidermal growth factor receptor mutants. Mol Cell Biol 12:491–498

Sato JD, Kawamoto T, Le AD, Mendelsohn J, Polikoff J, Sato GH (1983) Biological effects in vitro of monoclonal antibodies to human epidermal growth factor receptors. Mol Biol Med 1:511–529

Schechter AL, Stern DF, Vaidyanathan L, Decker SJ, Drebin JA, Greene MI, Weinberg RA (1984) The neu oncogene:an erb-B-related gene encoding a 185,000-Mr tumour antigen. Nature 312:513–516

Scott J, Urdea M, Quiroga M, Sanchez-Pescador R, Fong N, Selby M, Rutter WJ, Bell GI (1983) Structure of a mouse submaxillary messenger RNA encoding epidermal growth factor and seven related proteins. Science 221:236–240

Shaheen RM, Davis DW, Liu W, Zebrowski BK, Wilson MR, Bucana CD, McConkey DJ, McMahon G, Ellis LM (1999) Antiangiogenic therapy targeting the tyrosine kinase receptor for vascular endothelial growth factor receptor inhibits the growth of colon cancer liver metastasis and induces tumor and endothelial cell apoptosis. Cancer Res 59:5412–5416

Shalaby F, Rossant J, Yamaguchi TP, Gertsenstein M, WU XF, Breitman ML, Schuh AC (1995) Failure of blood-island formation and vasculogenesis in Flk-1-deficient mice. Nature 376:62–66

Sizeland AM, Burgess AW (1992) Anti-sense transforming growth factor alpha oligonucleotides inhibit autocrine stimulated proliferation of a colon carcinoma cell line. Mol Biol Cell 3:1235–1243

Slamon DJ, Clark GM, Wong SG, Levin WJ, Ullrich A, McGuire WL (1987) Human breast cancer:correlation of relapse and survival with amplification of the HER-2/neu oncogene. Science 235:177–182

Slamon DJ, Godolphin W, Jones LA, Holt JA, Wong SG, Keith DE, Levin WJ, Stuart SG, Udove J, Ullrich A, Press MF (1989) Studies of the HER-2/neu proto-oncogene in human breast and ovarian cancer. Science 244:707–712

Sures I, Goeddel DV, Gray A, Ullrich A (1980) Nucleotide sequence of human preproinsulin complementary DNA. Science 208:57–59

Terman BI, Dougher-Vermazen M, Carrion ME, Dimitrov D, Armellino DC, Gospodarowicz D, Bohlen P (1992) Identification of the KDR tyrosine kinase as a receptor for vascular endothelial cell growth factor. Biochem Biophys Res Commun 187:1579–1586

Ullrich A, Shine J, Chirgwin J, Pictet R, Tischer E, Rutter WJ, Goodman HM (1977) Rat insulin genes:construction of plasmids containing the coding sequences. Science 196:1313–1319

Ullrich A, Gray A, Berman C, Dull TJ (1983) Human beta-nerve growth factor gene sequence highly homologous to that of mouse. Nature 303:821–825

Ullrich A, Coussens L, Hayflick JS, Dull TJ, Gray A, Tam AW, Lee J, Yarden Y, Libermann TA, Schlessinger J, Downward J, Mayes ELV, Whittle N, Waterfield MD, Seeburg PH (1984) Human epidermal growth factor receptor cDNA sequence and aberrant expression of the amplified gene in A431 epidermoid carcinoma cells. Nature 309:418–425

Ullrich A, Bell JR, Chen EY, Herrera R, Petruzzelli LM, Dull TJ, Gray A, Coussens L, Liao YC, Rosen OM, Ramachandran J (1985) Human insulin receptor and its relationship to the tyrosine kinase family of oncogenes. Nature 313:756–761

Ullrich A, Gray A, Tam AW, Yang-Feng T, Tsubokawa M, Collins C, Henzel W, Le Bon T, Kathuria S, Chen E, Jacobs S, Francke U, Ramachandran J, Fujita-Yamaguchi Y (1986) Insulin-like growth factor I receptor primary structure:comparison with insulin receptor suggests structural determinants that define functional specificity. Embo J 5:2503–2512

Wakeling AE, Barker AJ, Davies DH, Brown DS, Green LR, Cartlidge SA, Woodburn JR (1996) Specific inhibition of epidermal growth factor receptor tyrosine kinase by 4-anilinoquinazolines. Breast Cancer Res Treat 38:67–73

Yaish P, Gazit A, Gilon C, Levitzki A (1988) Blocking of EGF-dependent cell proliferation by EGF receptor kinase inhibitors. Science 242:933–935

Yarden Y, Escobedo JA, Kuang WJ, Yang-Feng TL, Daniel TO, Tremble PM, Chen EY, Ando ME, Harkins RN, Francke U, Fried VA, Ullrich A, Williams LT (1986) Structure of the receptor for platelet-derived growth factor helps define a family of closely related growth factor receptors. Nature 323:226–232

Yarden Y, Kuang WJ, Yang-Feng T, Coussens L, Munemitsu S, Dull TJ, Chen E, Schlessinger J, Francke U, Ullrich A (1987) Human proto-oncogene c-kit: a new cell surface receptor tyrosine kinase for an unidentified ligand. Embo J 6:3341–51

Yamamoto T, Hihara H, Nishida T, Kawai S, Toyoshima K (1983) A new avian erythroblastosis virus, AEV-H, carries erbB gene responsible for the induction of both erythroblastosis and sarcomas. Cell 34:225–232

Targets for pituitary tumor therapy

Shlomo Melmed[1]

Summary

Recent advances in understanding the mechanisms underlying pituitary tumorigenesis will allow identification of novel targets for therapy for these common but often incurable tumors. Pituitary tumor initiation and progression are associated with multiple and acquired disorders. The pituitary gland responds to central and peripheral signals by undergoing reversible hormonal secretory changes and plastic cell growth changes. Underlying pituitary hyperplasia, with or without excess hormone production, or, in contrast, involution or hyposecretion in pituitary cells correlates with pituitary tumor development. Transgenic mouse models of tumor suppressor gene inactivation lead largely to the development of intermediate lobe tumors whereas pituitary-directed growth factor activation predisposes to anterior pituitary tumor development. Results of pituitary-directed pituitary tumor transforming gene (*PTTG*) inactivation or over-expression support the notion that the trophic environment is permissive for pituitary tumor formation. Understanding the mechanisms underlying pituitary plasticity and their relationship to tumor development will provide subcellular targets for treating both the development and growth of these tumors.

Introduction

Recent advances in understanding the pathogenesis of pituitary tumors have allowed several hypotheses to emerge regarding the primary etiology of these common adenomas. Although these tumors are invariably benign, long-term cure rates are difficult to achieve. Complex molecular cascades lead to pituitary adenoma formation and include factors regulating pituitary development, cell growth, and hormone gene expression. Extrinsic and intrinsic factors, including hypothalamic hormones, peripheral and local steroids, especially estrogens, as well as growth factor signals subserve pituitary gland plasticity, ranging from pituitary hypoplasia through hyperplasia, ultimately resulting in well-circumscribed hormone-secreting or non-secreting adenomas. These changes develop against a background of pituitary cell alterations, including chromosomal instability, epigenetic changes and mutations. Thus, a complex intrapituitary milieu potentiates the emergence of a monoclonal tumor cell population (Melmed 2003).

Descriptive studies of growth factor pituitary disruptions and activations have been extensively reviewed elsewhere (Musat et al. 2004; Farrell and Clayton 2003; Levy et al. 2003). Most descriptive reports do not provide validated predictors of clinical behavior

[1] Cedars-Sinai Medical Center, 8700 Beverly Blvd., Room 2015, Los Angeles, CA 90048

Conn et al.
Insights into Receptor Function
and New Drug Development Targets
© Springer-Verlag Berlin Heidelberg 2006

or of recurrence, and we hypothesize that pituitary cell trophic status is an important determinant of pituitary tumorigenesis.

Several animal models represent useful resources for evaluating determinants of pituitary plasticity as they relate to tumor formation. Most transgenic disruptions of tumor suppressor genes curiously result in the development of intermediate lobe tumors with or without POMC expression (Jacks et al. 1992; Kiyokawa et al. 1996). These melanocyte and POMC-related tumors may also be accompanied by concomitant anterior pituitary tumors expressing respective trophic hormones (Chesnokova et al. 2005; Nikitin et al. 1999). Transgenic overexpression of HMGA, involved in chromatin regulation (Fedele et al. 2002, 2005), galanin, a neuropeptide neurotransmitter (Perumal and Vrontakis 2003), and TGFα, a growth factor (McAndrew et al. 1995), results in high penetrance of anterior pituitary tumor development. These observations strongly indicate that a wide diversity of imbalances may trigger pituitary tumor initiation (Table 1). An abundance of pituitary tumor transforming gene (*PTTG*) correlates with changes in pituitary gland plasticity, i.e., transgenic models of *PTTG* inactivation result in pituitary hypoplasia whereas pituitary-directed PTTG overexpression results in hyperplasia and focal adenomas. Nevertheless, no activating *PTTG* mutation has yet emerged.

Table 1. A representative listing of known transgenic models for murine pituitary tumor development

	Hyperplasia/Adenoma	Reference
Gene overexpression[a]	Phenotype	
CMV.**HMGA 1**	GH, PRL	Fedele et al. 2005
CMV.**HMGA 2**	GH, PRL	Fedele et al. 2002
Ubiquitin C.**hCG**	PRL	Huhtaniemi et al. 2005
αGSU.**bLH**	Pit1 Lineage	Mohammad et al. 2003
GH.**galanin**	GH, PRL	Perumal et al. 2003
PRL.**galanin**	PRL [b]	Cai et al. 1999
PRL.**TGF**α	PRL	McAndrew et al. 1995
αGSU.**PTTG**	LH, GH, TSH	Abbud et al. 2005
Gene inactivation		
p27/*Kip1*−/−	ACTH, αMSH	Kiyokawa et al. 1996
p18/*INK4c*−/−	ACTH, αMSH	Bai et al. 2003
Rb+/−	ACTH,	Jacks et al. 1992
	αMSH αGSU, GH, βTSH	Nikitin et al. 1999
Men1+/−	PRL	Crabtree et al. 2001
PRL−/−	Non-secreting	Cruz-Soto et al. 2002

[a]Genes are listed in bold and are preceded by the promoter that determines transcriptional control
[b]Pituitary hyperplasia, with no tumor formation
CMV: cytomegalovirus; PRL: prolactin, Adapted from Donangelo and Melmed 2005

Pituitary plasticity and tumor development

Pituitary hypoplasia, with decreased cell number, is usually encountered in patients harboring genetic abnormalities with pituitary hormone deficiencies. Pituitary tumor development has not been encountered in these patients, and to date only a single case of corticotroph adenoma has been reported in a patient with disrupted *DAX-1* (De Menis et al. 2005). Pituitary tumors are frequently encountered at autopsy in up to 25% of cases or at incidental brain or nasal sinus MRI evaluation (Teramoto et al. 1994; Siquiera and Guembarovski 1984; Parent et al. 1981; Tomita and Gates 1999). Clinically active, i.e., hormone-secreting, pituitary tumors are less common, with a population prevalence of < 1% (Melmed and Kleinberg 2003). In 506 patients harboring incidental pituitary adenomas, the most frequent diagnosis was non-functioning pituitary adenoma (Sanno et al. 2003). Pituitary hyperplasia, however, is not commonly encountered in autopsy studies, likely due to diagnostic difficulties. Morphologic features and anatomical distribution differ between individual pituitary cell type hyperplasias, making this a very heterogeneous and diagnostically challenging pathological condition (Horvath et al. 1999).

Clinically, pituitary hyperplasia has rarely been documented to progress to neoplasia. New pituitary adenomas are uncommon in patients with physiological hyperplasia associated with pregnancy or lactation, or pituitary enlargement due to estrogen administration, or longstanding primary hypothyroidism (Horvath et al. 1999; Coogan et al. 1995; Kovacs et al. 1994; Ghannam et al. 1999). Patients with GHRH (Growth Hormone Releasing Hormone) producing tumors develop acromegaly due to somatotroph hyperplasia (Sano et al. 1988), and only rarely associated with a growth hormone (GH)-secreting adenoma (Shintani et al. 1995; Sano et al. 1988). Pituitary adenomas are monoclonal in origin and transformed cells are usually surrounded by non-hyperplastic pituitary tissue (Clayton and Farrell 2004). In contrast, sustained long-term pituitary hyperplasia in rodents ultimately results in adenoma progression (Heaney et al. 1999a; Asa et al. 1992). Prospectively, it is important to determine whether or not sustained (over years) pituitary hyperplasia results in a higher incidence of adenoma development in patients with unambiguously demonstrated pituitary hyperplasia.

PTTG and pituitary gland plasticity

PTTG isolated from pituitary tumor cells results in cellular transformation and tumor formation in nude mice (Pei and Melmed 1997). PTTG, the index mammalian securin protein, regulates sister chromatid separation during mitosis (Zou et al. 1999), and excess or low PTTG levels result in cell aneuploidy (Yu et al. 2003; Wang et al. 2001). Levels are elevated in pituitary (Zhang et al. 1999), thyroid (Heaney et al. 2001), colon (Heaney et al. 2000), and breast (Solbach et al. 2004) tumors, especially when invasive. Mechanisms accounting for the transforming role of *PTTG* include 1) chromosome instability, 2) induction of growth factors (Heaney et al. 1999b), and 3) transactivation of oncogenes or cell cycle proteins (Pei 2001).

Transgenic *PTTG1* expression driven by the α-subunit glycoprotein (αGSU) promoter in mice results in PTTG co-expression in LH-, FSH-, TSH-secreting cells, and also in GH-producing cells (Abbud et al. 2005). αGSU.PTTG mice develop a broad range of pituitary changes, including pituitary plurihormonal hyperplasia to hormone-secreting microadenomas (Fig. 1). These hormonally active tumors result in endocrine phenotypes including high IGF-1 and testosterone levels leading to prostate and seminal vesicle enlargement. As only small microadenomas arise in a subset of these

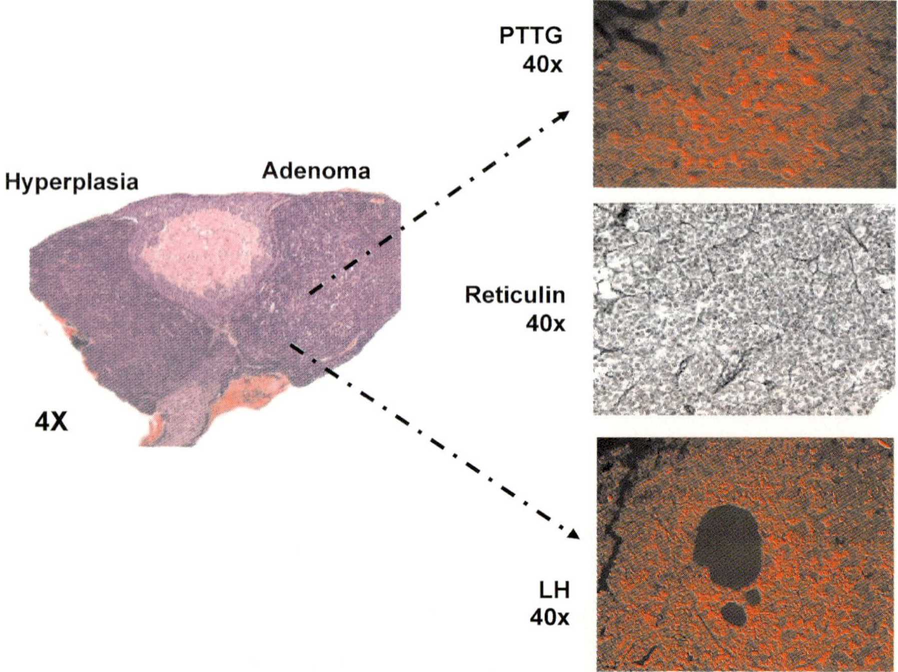

Fig. 1. Pituitary PTTG expression in αGSU.PTTG mice results in hyperplasia and adenomatous foci. *Left*: Representative pituitary gland harboring a small adenoma. *Right*: Breakdown in reticulin network confirms the diagnosis of adenoma vs. hyperplasia. Adenomas are highly immunoreactive for PTTG, and typically stain positive for LH or GH. (Adapted from Abbud et al. 2005, with permission)

animals, despite the very high penetrance of pituitary hyperplasia, additional factors are likely required for macroadenoma development. To study the interaction of *PTTG* with other cell cycle regulators, αGSU.PTTG mice were cross-bred with mice heterozygous for *Rb1* inactivation (Rb1+/−). The latter animals develop intermediate pituitary lobe tumors with high penetrance (Jacks et al. 1992; Hu et al. 1994) and, less commonly, anterior lobe pituitary adenomas (Nikitin et al. 1999). The compound effects of pituitary *PTTG* overexpression together with Rb1 inactivation result in additive pituitary enlargement, suggesting that *PTTG* and *Rb1* play opposing roles in impeding pituitary tumor development.

In contrast, *Pttg* inactivation restrains pituitary growth and tumorigenic capacity. *Pttg* knockout results in pituitary, pancreatic β-cell, spleen, and testicular hypoplasia (Wang et al. 2001). Embryonic fibroblasts derived from *Pttg* null mice exhibit premature

Fig. 2. *Pttg* deletion suppresses pituitary tumor development in $Rb^{+/-}$ mice. Development of pituitary tumors in *wt*, $Rb^{+/-}$, $Pttg^{-/-}$ and $Rb^{+/-}Pttg^{-/-}$ mice over time: Rb+/− mice have a cumulative pituitary tumor incidence of 86% by 13 months, whereas 20% of Pttg−/−Rb+/− mice develop pituitary tumors at the same age ($p < 0.01$); ~15% of WT and Pttg−/− mice develop pituitary tumors at a late age. (Reproduced from Chesnokova et al. (2005), with permission)

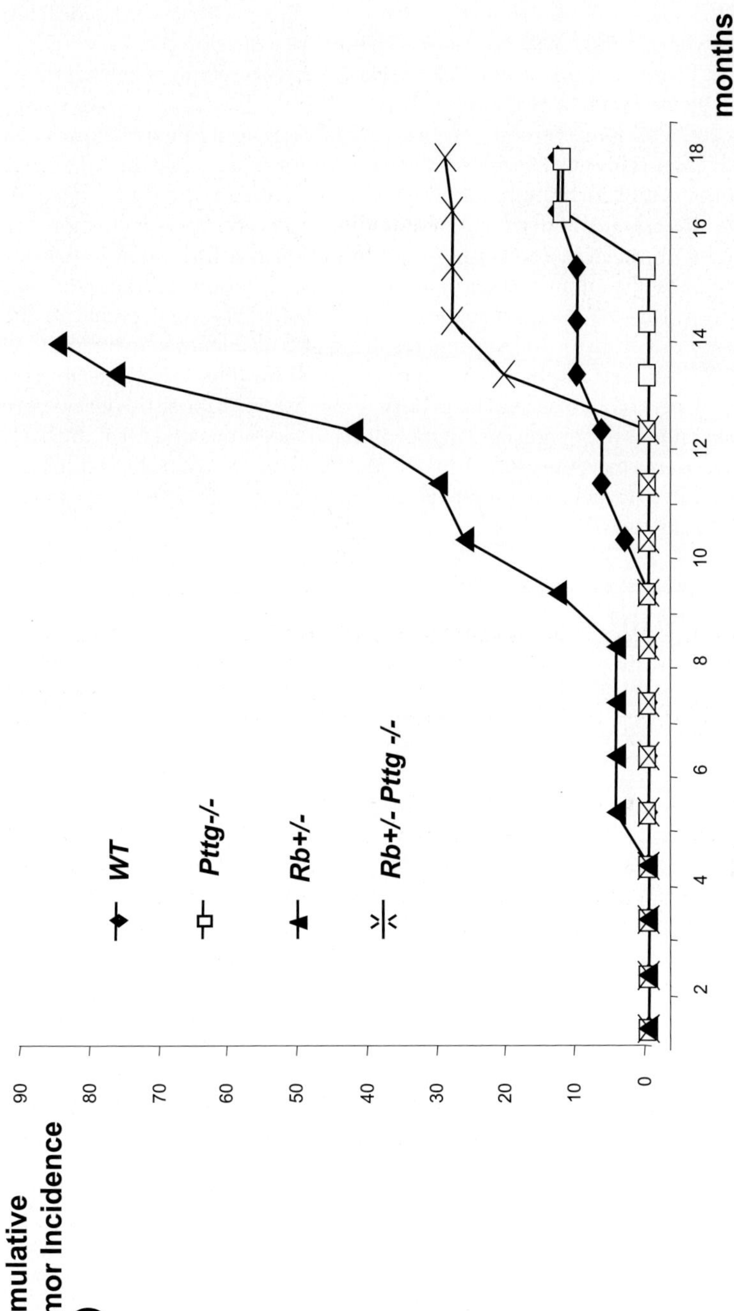

Cumulative Tumor Incidence (%)
months
WT
Pttg-/-
Rb+/-
Rb+/- Pttg -/-

centromere separation, chromosome mis-segregation, and aneuploidy (Wang et al. 2001). Despite causing aneuploidy, low *Pttg* levels restrain and delay pituitary tumor development (Chesnokova et al. 2005). By 13 months, 86% of Rb+/− mice develop pituitary tumors whereas only 30% of Pttg−/−Rb+/− mice develop pituitary tumors ($P < 0.01$) (Fig. 2). Pttg also suppresses P21 levels, and *Pttg* inactivation likely restrains tumor formation by up-regulating p21 expression.

In summary, low Pttg levels result in pituitary hypoplasia, and *Pttg* overexpression results in pituitary hyperplasia. These examples of pituitary plasticity directly relate to the potential for tumor formation. Pathways that induce PTTG and ultimately in tumor formation are as yet not apparent, although estrogen receptor activation has been shown to induce both PTTG levels and tumor initiation in animal models (Heaney et al. 1999). In the absence of a demonstrated activating mutation, current evidence supports the notion that *PTTG* perturbations are probably proximal events in the cascade of pituitary cell-transforming events resulting in true adenoma formation. As depicted in Fig. 3, pituitary trophic status appears to be an important determinant for the monoclonal adenoma to arise. The genotype-phenotype models reviewed here represent the spectrum of hypopolasia → hyperplasia → adenoma and form the basis for further testing of this hypothesis in humans. Mechanisms for controlling pituitary plasticity provide novel targets for therapeutics that control both the initiation and progression of pituitary tumors.

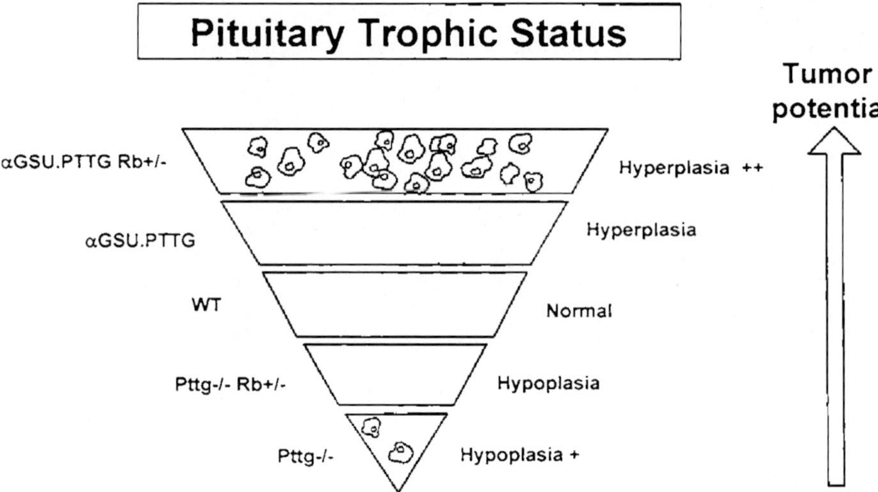

Fig. 3. Cartoon depiction of trophic determinants of pituitary tumor formation

References

Abbud RA, Takumi I, Barker EM, Ren SG, Chen DY, Wawrowsky K, Melmed S (2005) Early multipotential pituitary focal hyperplasic in αGSU-driven pituitary tumor transforming gene (PTTG) transgenic mice. Mol Endocrinol 19:1383–1391

Asa SL, Kovacs K, Stefaneanu L, Horvath E, Billestrup N, Gonzalez-Manchon C, Vale W (1992) Pituitary adenomas in mice transgenic for growth hormone-releasing hormone. Endocrinology 131:2083–2089

Bai F, Pei XH, Godfrey VL, Xiong Y (2003) Haploinsufficiency of p18(INK4c) sensitizes mice to carcinogen-induced tumorigenesis. Mol Cell Biol 23:1269–1277

Cai A, Hayes JD, Patel N, Hyde JF (1999) Targeted overexpression of galanin in lactotrophs of transgenic mice induces hyperprolactinemia and pituitary hyperplasia. Endocrinology 140:4955–4964

Chesnokova V, Kovacs K, Castro AV, Zonis S, Melmed S (2005) Pituitary hypoplasia in Pttg−/− mice is protective for Rb+/− pituitary tumorigenesis. Mol Endocrinol 19:2371–2379

Clayton RN, Farrell WE (2004) Pituitary tumour clonality revisited. Front Horm Res 32:186–204

Coogan PF, Baron JA, Lambe M (1995) Parity and pituitary adenoma risk. J Natl Cancer Inst 87:1410–1411

Crabtree JS, Scacheri PC, Ward JM, McNally SR, Swain GP, Montagna C, Hager JH, Hanahan D, Edlund H, Magnuson MA, Garrett-Beal L, Burns AL, Ried T (2001) A mouse model of multiple endocrine neoplasia, type 1, develops multiple endocrine tumors. Proc Natl Acad Sci USA 98:1118–1123

Cruz-Soto ME, Scheiber MD, Gregerson KA, Boivin GP, Horseman ND (2002) Pituitary tumorigenesis in prolactin gene-disrupted mice. Endocrinology 143:4429–4436

Cushman LJ, Burrows HL, Seasholtz AF, Lewandoski M, Muzyczka N, Camper SA (2000) Cre-mediated recombination in the pituitary gland. Genesis 28:167–174

De Menis E, Roncaroli F, Calvari V, Chiarini V, Pauletto P, Camerino G, Cremonini N (2005) Corticotroph adenoma of the pituitary in a patient with X-linked adrenal hypoplasia congenita due to a novel mutation of the DAX-1 gene. Eur J Endocrinol 153:211–215

Donangelo I, Melmed S (2005) Pathophysiology of pituitary adenomas. J Endocrinol Invest 28 (Suppl 11):102–107

Farrell WE, Clayton RN (2003) Epigenetic change in pituitary tumorigenesis. Endocr Relat Cancer 10:323–330

Fedele M, Battista S, Kenyon L, Baldassarre G, Fidanza V, Klein-Szanto AJ, Parlow AF, Visone R, Pierantoni GM, Outwater E, Santoro M, Croce CM, Fusco A (2002) Overexpression of the HMGA2 gene in transgenic mice leads to the onset of pituitary adenomas. Oncogene 21:3190–3198

Fedele M, Pentimalli F, Baldassarre G, Battista S, Klein-Szanto AJ, Kenyon L, Visone R, De Martino I, Ciarmiello A, Arra C, Viglietto G, Croce CM, Fusco A (2005) Transgenic mice overexpressing the wild-type form of the HMGA1 gene develop mixed growth hormone/prolactin cell pituitary adenomas and natural killer cell lymphomas. Oncogene 24:3427–3435

Ghannam NN, Hammami MM, Muttair Z, Bakheet SM (1999) Primary hypothyroidism-associated TSH-secreting pituitary adenoma/hyperplasia presenting as a bleeding nasal mass and extremely elevated TSH level. J Endocrinol Invest 22:419–423

Heaney AP, Horwitz GA, Wang Z, Singson R, Melmed S (1999a) Early involvement of estrogen-induced pituitary tumor transforming gene and fibroblast growth factor expression in prolactinoma pathogenesis. Nature Med 5:1317–1321

Heaney AP, Horwitz GA, Wang Z, Singson R, Melmed S (1999b) Early involvement of estrogen-induced pituitary tumor transforming gene and fibroblast growth factor expression in prolactinoma pathogenesis. Nature Med 5:1317–1321

Heaney AP, Singson R, McCabe CJ, Nelson V, Nakashima M, Melmed S (2000) Expression of pituitary-tumour transforming gene in colorectal tumours. Lancet 355:716–719

Heaney AP, Nelson V, Fernando M, Horwitz G (2001) Transforming events in thyroid tumorigenesis and their association with follicular lesions. J Clin Endocrinol Metab 86:5025–5032

Horvath E, Kovacs K, Scheithauer BW (1999) Pituitary hyperplasia. Pituitary 1:169–179

Hu N, Gutsmann A, Herbert DC, Bradley A, Lee W-H, Lee EY-HP (1994) Heterozygous Rb-$1^{\Delta 20}$/+mice are predisposed to tumors of the pituitary gland with a nearly complete penetrance. Oncogene 9:1021–1027

Huhtaniemi I, Rulli S, Ahtiainen P, Poutanen M (2005) Multiple sites of tumorigenesis in transgenic mice overproducing hCG. Mol Cell Endocrinol 234:117–126

Jacks T, Fazeli A, Schimitt EM, Bronson RT, Goodell MA, Weinberg RA (1992) Effects of an Rb mutation in the mouse. Nature 359:295–300

Kiyokawa H, Kineman RD, Manova-Todorova KO, Soares VC, Hoffman ES, Ono M, Khanam D, Hayday AC, Frohman LA, Koff A (1996) Enhanced growth of mice lacking the cyclin-dependent kinase inhibitor function of p27(Kip1). Cell 85:721–732

Kovacs K, Stefaneanu L, Ezzat S, Smyth HS (1994) Primary hypothyroidism-associated TSH-secreting pituitary adenoma/hyperplasia presenting as a bleeding nasal mass and extremely elevated TSH level. Arch Pathol Lab Med 118:562–565

Levy A, Lightman S (2003) Molecular defects in the pathogenesis of pituitary tumours. Front Neuroendocrinol 24, 94–127

McAndrew J, Paterson AJ, Asa SL, McCarthy KJ, Kudlow JE (1995) Targeting of transforming growth factor-alpha expression to pituitary lactotrophs in transgenic mice results in selective lactotroph proliferation and adenomas. Endocrinology 136:4479–4488

Melmed S (2003) Mechanisms for pituitary tumorigenesis: the plastic pituitary. J Clin Invest 112:1603–1618

Melmed S, Kleinberg K (2003) Anterior pituitary. In: Larsen PR, Kronenberg HM, Melmed S, Polonsky KS (eds) Williams' textbook of endocrinology 10th ed. Philadelphia Elsevier, p 177–280

Mohammad HP, Abbud RA, Parlow AF, Lewin JS, Nilson JH (2003) Targeted overexpression of luteinizing hormone causes ovary-dependent functional adenomas restricted to cells of the Pit-1 lineage. Endocrinology 144:4626–4636

Musat M, Vax VV, Borboli N, Gueorguiev M, Bonner S, Korbonits M, Grossman AB (2004) Cell cycle dysregulation in pituitary oncogenesis. Front Horm Res 32:34–62

Nikitin AY, Juárez-Pérez MI, Li S, Huang L, Lee W-H (1999) RB-mediated suppression of spontaneous multiple neuroendocrine neoplasia and lung metastases in Rb+/− mice. Proc Natl Acad Aci USA 96:3916–3921

Parent AD, Bebin J, Smith RR (1981) Incidental pituitary adenomas. J Neurosurg 54:228–231

Pei L (2001) Identification of c-myc as a down-stream target for pituitary tumor-transforming gene. J Biol Chem 276:8484–8491

Pei L, Melmed S (1997) Isolation and characterization of a pituitary tumor-transforming gene (PTTG). Mol Endocrinol 11:433–441

Perumal P, Vrontakis ME (2003) Transgenic mice over-expressing galanin exhibit pituitary adenomas and increased secretion of galanin, prolactin and growth hormone. J Endocrinol 179:145–154

Sanno N, Oyama K, Tahara S, Teramoto A, Kato Y (2003) A survey of pituitary incidentaloma in Japan. Eur J Endocrinol 149:123–127

Sano T, Asa SL, Kovacs K (1988) Growth hormone releasing hormone producing tumors: clinical, biochemical and morphological manifestations. Endocr Rev 9:357–373(6)

Shintani Y, Yoshimoto K, Horie H, Kanesaki Y, Hosoi E, Yokogoshi Y, Bando H, Iwahana H, Kannuki S, Matsumoto K, Itakura M, Saito S (1995) Two different pituitary adenomas in a patient with multiple endocrine neoplasia type. 1 associated with growth hormone-releasing hormone-producing pancreatic tumor: clinical and genetic features. Endocr J 42:331–340

Siqueira MG, Guembarovski AL (1984) Subclinical pituitary microadenomas. Surg Neurol 22:134–140

Solbach C, Roller M, Fellbaum C, Nicoletti M, Kaufmann M (2004) PTTG mRNA expression in primary breast cancer: a prognostic marker for lymph node invasion and tumor recurrence. Breast 13:80–81

Teramoto A, Hirakawa K, Sanno N, Osamura Y (1994) Incidental pituitary lesions in 1,000 unselected autopsy specimens. Radiology 193:161–164

Tomita T, Gates E (1999) Pituitary adenomas and granular cell tumors. Incidence, cell type, and location of tumor in 100 pituitary glands at autopsy. Am J Clin Pathol 111:817–825

Wang Z, Yu R, Melmed S (2001) Mice lacking pituitary tumor transforming gene show testicular and splenic hypoplasia, thymic hyperplasia, thrombocytopenia, aberrant cell cycle progression, and premature centromere division. Mol Endocrinol 15:1870–1879
Yu R, Lu W, Chen J, McCabe CJ, Melmed S (2003) Overexpressed pituitary tumor-transforming gene causes aneuploidy in live human cells. Endocrinology 144:4991–4998
Zhang X, Horwitz GA, Heaney AP, Nakashima M, Prezant TR, Bronstein MD, Melmed S (1999) Pituitary tumor transforming gene (PTTG) expression in pituitary adenomas. J Clin Endocrinol Metab 84:761–767
Zou H, McGarry TJ, Bernal T, Kirschner MW (1999) Identification of a vertebrate sister-chromatid separation inhibitor involved in transformation and tumorigenesis. Science 285:418–422

plant *Cannabis* are a group of C_{21} monoterpenoid derivatives, named cannabinoids. About 70 naturally occurring cannabinoids are known today, but the most important representative of these is Δ^9-tetrahydrocannabinol (Δ^9-THC), which has psychotropic properties and is responsible for many of the pharmacological actions of *Cannabis* (Nocerino et al. 2000).

The molecular targets of Δ^9-THC are at least two types of receptors, i.e., CB_1 and CB_2 receptors, both of which are coupled to $G_{i/o}$ proteins (Giuffrida et al. 2001; Howlett et al. 2002). CB_1 receptors (identified pharmacologically in 1988 and cloned in 1990; Devane et al. 1988); Matsuda et al. 1990) are expressed mostly by central and peripheral neurons, whereas CB_2 receptors (cloned in 1993; Munro et al. 1993) are expressed mostly by immune cells. The discovery of cannabinoid receptors was followed in 1992 by the demonstration that anandamide (arachidonyl ethanolamine) is an endogenous ligand (endocannabinoid) for these receptors (Devane et al. 1992). Other endocannabinoids identified are 2-arachidonoyl glycerol (2-AG), isolated in 1995 (Mechoulam et al. 1995; Sugiura et al. 1995) and noladin ether (2-arachidonyl glyceryl ether), isolated in 2001 (Hanus et al. 2001). Anandamide is also an endogenous ligand for vanilloid VR_1 receptors (also named capsaicin receptors; Zygmunt et al. 1999), the molecular target of chili.

Endocannabinoids have been identified as retrograde signalling molecules in the brain (Wilson and Nicoll 2002). The synthesis of ndocannabinoids from membrane lipid precursors is triggered by calcium influx into postsynaptic cells. Endocannabinoids then leave the postsynaptic cell and activate presynaptic CB_1 receptors, resulting in inhibition of neurotransmitter release. After leaving the receptor, endocannabinoids are removed from the extracellular space by a carrier (anandamide membrane transporter, AMT)-mediated, saturable uptake process (Fowler and Jacobsson 2002). Within the cell, anandamide is hydrolyzed to arachidonic acid and ethanolamine by fatty acid amide hydrolase (FAAH, also called anandamide amidohydrolase; Deutsch et al. 2002). FAAH can also catalyze the hydrolysis of 2-AG, an indication that it has esterase as well as amidase activity (Sugiura et al. 2002). Cannabinoid receptors, their endogenous ligands (endocannabinoids) and the proteins participating in the inactivation of these compounds are components of the so-called "endogenous cannabinoid system". Drugs able to affect the activity of the endogenous cannabinoid system include cannabinoid receptor agonists, inhibitors of endocannabinoids inactivation (indirect agonists such as FAAH and/or AMT inhibitors) and cannabinoid receptor antagonists

Although a pharmacological manipulation of the endogenous cannabinoid system could have potential therapeutic applications in the treatment of pain, neurodegenerative and musculoskeletal disorders, liver cirrhosis, glaucoma, inflammation and cancer (Nocerino et al. 2000), this review will deal with the role of the endogenous cannabinoid system in eating behavior and in the control of food intake, with a special focus on the potential therapeutic applications of cannabinoid CB_1 antagonists in the treatment of obesity.

Cannabis and human appetite

Cannabis has been used since antiquity for the treatment of many ailments, including eating disorders. For example, marijuana was recommended in India to treat loss of appetite in 300 AD (Touw 1981). William O'Shaughnessy played a leading role in introducing this substance to Western medicine in the middle of the 18th century. One of the many pharmacological properties of *Cannabis* he described was the induction of a "remarkable increase of appetite" (Abel 1975). During the 19th century, physicians mentioned the increased stimulation of appetite following *Cannabis* use (Abel 1975). Thus, in 1845, Donovan reported that *Cannabis* was effective in various inflammatory diseases, and he observed its effect on hunger. Birch, in 1899, reported from Calcutta that *Cannabis* was valuable in the treatment of opium addiction and that it "restored the ability to appreciate food" (Berry and Mechoulam 2002).

The first protocol committed to study the effect of marijuana on food intake was performed by the military in 1933 (Siler et al. 1933). Soldiers smoking marijuana were described to feel hungry, eating much more than control subjects. The first controlled study of the effect of marijuana on food intake was performed in 1971. This study showed that the acute effect of oral doses of *Cannabis* extract standardized to Δ^9-THC, administered to young volunteers in fasted or fed conditions, increased food intake as compared to placebo. However, the effect was statistically significant only in fed subjects (Hollister 1971). In another controlled study, a clear confirmation of the increased desire for food (marshmallows) in adult volunteer subjects smoking marijuana was found (Abel 1971). Importantly, chronic treatments also demonstrated that smoked marijuana significantly increased mean daily caloric intake (Greenberg et al. 1976; Foltin et al. 1986) and, specifically, increased the consumption of sweets (Foltin et al. 1988).

This stimulating effect on appetite observed in healthy subjects suggested the utility of cannabinoids in treatment of clinical syndromes that featured appetite or weight loss, such as cancer or AIDS-associated anorexia. In 1970, a placebo-controlled clinical study showed that oral Δ^9-THC significantly increased weight gain in patients with advanced cancer (Regelson et al. 1976). In 1992, Δ^9-THC (dronabinol) was approved by the Food and Drug Administration (FDA) in the USA for the treatment of patients with HIV-induced wasting syndrome. In all studies of dronabinol administration in AIDS patients, a noticeable stimulating effect of the drug on appetite was demonstrated (for a review, see Cota et al. 2003a).

Cannabinoids and feeding in animal studies

Animal studies have confirmed the anecdotal data in humans regarding the effect of *Cannabis* on hunger. An early study showed an increase in food intake in rats after the intraperitoneal administration of *Cannabis* (Carlini and Kramer 1965). In the following years, a number of experimental studies identified Δ^9-THC as the main active ingredient responsible for *Cannabis*-induced food intake in rodents (Abel 1975). It is important to note that not all of the studies yielded positive results (i.e., an increase in food intake). This was due to the fact that high doses (more than 10 mg/kg) of Δ^9-THC were used in most studies, and doses above 10 mg/kg Δ^9-THC are known to produce sedating effects in animals.

In 1977, a study by Brown and colleagues suggested that low doses of Δ^9-THC produced a dose- and time-dependent preference towards food and sucrose solution intake in rats (Brown et al. 1977). This study was the first to corroborate human anecdotal data regarding the notion that cannabinoids exert a preferential effect on palatable foods.

Insights into the mechanism of action of cannabinoid-induced hyperphagia were provided by the discovery of the cannabinoid receptors and with the development of cannabinoid receptor agonists and antagonists. Administration of cannabinoid drugs in animals revealed that the hyperphagic effect induced by Δ^9-THC was mediated by activation of CB_1 receptors. Importantly, it was shown that CB_1 receptor antagonists not only antagonized the appetite-stimulating effect of cannabinoid agonists (Williams et al. 1998, 1999; Freedland et al. 2000) but also had an action opposite to that of cannabinoid receptor agonists when administered alone (i.e., CB_1 receptor antagonists, per se, reduced food intake). For example, intraperitoneal injection of the CB_1 receptor antagonist rimonabant suppressed appetite and induced weight loss in rats (Arnone et al. 1997; Colombo et al. 1998). Moreover, rimonabant was reported to selectively reduce sweet food intake in marmosets (Simiand et al. 1998). These studies can be considered as providing the experimental basis for the possible clinical application of CB_1 antagonists in the treatment of obesity (Fernandez and Allison, 2004; Carai et al. 2005; Boyd and Fremming 2005).

The availability of CB_1 receptor knockout mice has provided important supporting evidence for endocannabinoid involvement in appetite regulation. When maintained on standard chow, CB_1 receptor knockout mice were leaner and slightly hypophagic compared to wild-type mice animals. When fed a palatable, high-fat diet, CB_1 receptor knockout mice did not display the hyperphagia characteristic of wild-type mice and did not develop obesity (Ravinet-Trillou et al. 2003). Additionally, CB_1 receptor knockout mice showed reduced consumption of sucrose compared to wild type mice (Poncelet et al. 2003). Finally, Di Marzo and colleagues showed that CB_1 receptor knockout mice displayed a reduced hyperphagic response to fasting, eating less than wild-type littermates (Di Marzo et al. 2001).

Site of action of cannabinoids

The use of animal models also represents a potential tool for understanding the mechanism and the site of action of *Cannabis* and its constituents. Recent evidence suggests that the endogenous cannabinoid system may regulate energy balance and food intake at several functional levels, both in the brain (hypothalamus, limbic system) and in peripheral organs involved in energy storage and expenditure. Hypothalamus, limbic system, gastrointestinal tract, adipose tissue, liver and skeletal muscle are possible sites of action of cannabinoid drugs.

Hypothalamus

CB_1 receptors and endocannabinoids are present in the hypothalamic areas involved in food intake, and they are able to cross-talk to signals of peripheral origin (for reviews, see Fride 2004, 2005; Pagotto et al. 2005). It is believed that the endogenous cannabinoid system is activated "on demand" in the hypothalamus after short-term food deprivation and then transiently regulates the levels and/or action of other orexigenic and anorectic mediators to induce appetite (Di Marzo and Matias 2005).

Experimental studies have shown that injection of anandamide into the ventromedial hypothalamus stimulates appetite in rats (Jamshidi and Taylor 2001). Moreover, 24 hours of food deprivation produces an increase in the hypothalamic levels of 2-AG in rats (Kirkham et al. 2002) and mice (Hanus et al. 2003). The levels of 2-AG decline after eating and return to control levels with the onset of satiety. These changes seem to be inversely correlated with the changes that are known to occur in blood levels of the neurohormone leptin, which is pivotal in regulating the hypothalamic orexigenic and anorectic signals. Indeed, leptin decreases endocannabinoid levels in the hypothalamus, and obese rodents with defective leptin signalling show significantly higher hypothalamic endocannabinoid levels (Di Marzo et al. 2001).

The endocannabinoid system may influence food intake by regulating the expression and/or action of several hypothalamic anorectic and orexigenic mediators (Di Marzo and Matias 2005). CB_1 receptors colocalize with corticotrophin-releasing hormone (CRH) in the paraventricular nucleus, with melanin-concentrating hormone (MCH) in the lateral hypothalamus, and with pre-pro-orexin in the ventromedial hypothalamus (Cota et al. 2003a). There is also evidence for functional interaction between endocannabinoids and orexin A, an orexigenic peptide that has been linked to the stimulation of feeding. Indeed, evidence for hypersensitization of the orexin 1 receptor by the CB_1 receptor (cross-talk) has been reported (Hilairet et al. 2003). Additionally, CB_1 receptor knockout mice (which are characterized by hypophagia, reduced body weight, and reduced fat mass compared with their wild-type littermates) show higher levels of RNA for the anorexigen CRH (Cota et al. 2003b).

Limbic system

Cannabinoid could partly modify food intake by hedonic response to foods (Fride 2004; Vickers and Kennett, 2005). Administration of the endocannabinoid 2-AG into the nucleus accumbens (specifically into the shell subregion of the nucleus accumbens, which is involved in the generation of emotional arousal and behavioral activation in response to rewarding stimuli) evokes a dramatic hyperphagic response (Kirkham et al. 2002). Moreover, fasting increases levels of both anandamide and 2-AG in the limbic forebrain (Kirkham et al. 2002), whereas over-consumption of palatable foods down-regulates CB_1 receptor expression in the nucleus accumbens.

Additional evidence suggests the existence of a relationship between endocannabinoid and opioids in mediating the rewarding properties of foods (Kirkham and Williams 2004). Indeed 1) the hyperphagic effect of Δ^9-THC rats is counteracted by the opioid receptor antagonist naloxone (Kirkham 1990), 2) doses of naloxone and rimonabant, which are per se inactive, synergistically produce an anorectic action when co-administered (Kirkham and Williams, 2001; Rowland et al. 2001; Chen et al. 2004) and 3) feeding induced by injection of morphine into the paraventricular nucleus within the hypothalamus can be reversed by rimonabant (Tucci et al. 2004).

Gastrointestinal

The digestive tract contains endogenous cannabinoids (anandamide and 2-AG), and cannabinoid CB_1 receptors can be found on myenteric and submucosal nerves. Activation of CB_1 receptors produces inhibition of gastrointestinal motility and secretion as well as antinflammatory effects (Coutts and Izzo 2004). Recent evidence suggests that the intestinal endocannabinoid system might be involved in the control of appetite and this action might be modulated by peripheral peptides [e.g., cholecystokinin (CCK)], which are known to have a role in feeding.

The endogenous cannabinoid system in the gut undergoes adaptive changes in response to diet. Food deprivation produced a seven-fold increase in anandamide content in the small intestine but not in the brain or stomach. This effect was associated with increased expression of vagal CB_1 receptors (Burdyga et al. 2004). Refeeding normalized intestinal anandamide levels and CB_1 expression. Moreover, capsaicin deafferentation abolished the hyperphagic action of the cannabinoid agonist as well as the anorectic action of rimonabant (Gomez et al. 2002). These findings suggest that CB_1 receptors, located on capsaicin-sensitive sensory neurons, may be involved in cannabinoid-induced modulation of appetite and that anandamide acts as a "hunger signal".

CCK is an anorexigenic peptide produced in gall bladder, pancreas, and stomach and concentrated in the small intestine. CCK receptor agonism inhibits gastric emptying and primarily increases central signalling of satiety through vagal afferent signals to the brain, resulting in short-term inhibition of food intake (Bays 2004). It is therefore noteworthy that gastric and intestinal vagal afferents that express CCK receptors also express CB_1 receptors. CCK administration reduces food intake and also decreases CB_1 receptor expression in vagal afferent neurons. These data suggest that the endocannabinoid system may influence food intake by a gastroduodenal cross-talk with CCK signalling (Burdyga et al. 2004).

Adipose tissue

The major known metabolic functions of adipose tissue include uptake of circulating nutrients, lypolysis of stored triglycerides, removal of circulating lipoprotein-triglycerides by the action of lipoprotein lipases and synthesis and storage of triglycerides (lipogenesis; Di Girolamo et al. 2000).

Cota and coworkers (2003b) showed that the CB_1 receptor is functionally active in white adipocytes stimulating lipogenesis. CB_1 activation enhances lipoprotein lipase activity and this effect has been shown to be specifically blocked by rimonabant (Cota et al. 2003b). In addition, Bensaid and coworkers (2003) showed that CB_1 receptor is mostly expressed in mature adipocytes when compared to preadipocytes and that CB_1 expression is increased in adipocytes from obese mice compared to those derived from lean ones. These facts raise the possibility that the endogenous cannabinoid system may influence body weight with a mechanism not involving food intake. Indeed, the CB_1 receptor antagonist rimonabant has been shown to reduce adiposity in diet-induced obese mice and genetically obese rodents, independently of its anorectic action (Hildebrant) et al. 2003; Ravinet-Trillou et al. 2003). Thus, while chronic CB_1 blockade initially suppresses food intake, this action is seen to gradually wane. In contrast, weight loss persists even after the marked anorectic effects of the antagonists. CB_1 receptor blockade may, therefore, interfere with the processes that regulate fat deposition in adipose tissue and/or may enhance fatty acid oxidation, since rimonabant is known to lower plasma-free fatty acid levels in dietary obese mice (Ravinet-Trillou et al. 2003). Importantly, rimonabant induces adiponectin release from adipocytes in vitro (Bensaid et al. 2003). Adiponectin is an adipose tissue-specific protein that plays a key regulatory role in fat and glucose metabolism (adiponectin is associated with an increased insulin sensitivity, and in the liver it decreases hepatic glucose production and regulates free fatty acid metabolism, via suppression of lipogenesis and activation of free fatty acid oxidation).

Liver

As a major site of lipogenesis, the liver has also received attention as a possible site of cannabinoid activity (Lichtman and Cravatt, 2005). Recently, Osei-Hyiaman and colleagues (2005) have shown that obesity induced by over-consumption of fat leads to elevated anandamide levels in the liver (through reduction of its enzymatic breakdown), up-regulation of hepatic CB_1 receptors, and increased hepatic fatty acid synthesis. These changes are prevented by the CB_1 receptor antagonist rimonabant and are absent in CB_1 knockout mice. Systemic administration of the potent cannabinoid receptor agonist HU210 led to significant increases in mRNA and protein levels of SREBP-1c, as well as those of its target enzymes, acetyl-CoA carboxylase-1 (ACC1) and fatty acid synthase (FAS). A functional consequence of the activation of these proteins was a two-fold increase in the rate of fatty acid synthesis in liver, an effect that was blocked by the CB1 receptor antagonist rimonabant and did not occur in CB_1 knockout mice. HU210 also increased SREBP-1c and FAS mRNA levels in the hypothalamus through a CB_1-mediated mechanism (Osei-Hyiaman et al. 2005). Moreover, the observation that CB_1 knockout mice and CB_1 wild-type mice had a similar total caloric intake, independent of diet, supports the notion that increased cannabinoid-mediated lipogenesis plays a more dominant role in diet-induced obesity.

These authors also showed that wild-type mice maintained on a high-fat diet gained significant amounts of weight, mostly in the form of adipose tissue, had altered levels of enzymes associated with metabolism (insulin, leptin, and adiponectin), had an increased rate of hepatic fatty acid synthesis, developed fatty liver and showed increased levels of triglycerides. In contrast, CB_1 knockout mice maintained on a high-fat died did not exhibit an increase in fatty acid synthesis, exhibited serum hormone and lipid profiles similar to wild-type mice fed regular chow, and did not develop fatty liver (Osei-Hyiaman et al. 2005).

In summary, activation of CB_1 receptor in the liver plays a key role in increased serum lipid production, fatty liver, and possibly diet-induced obesity. Conversely, stimulation of these receptors in the hypothalamus may lead to an increase in food consumption. Thus, targeting both of these pathways with CB_1 antagonists could promote sustained weight loss and favorable serum lipid profiles in obese patients (Lichtman and Cravatt 2005).

Skeletal muscle

In adult humans, the major thermogenetic tissue is skeletal muscle, which, in non-obese subjects, comprises approximately 40% of body weight and accounts for 20–30% of the total oxygen consumption at rest (Chiesi et al. 2001). Very recently Liu and colleagues (2005) have investigated the effect of the CB_1 receptor antagonist rimonabant on energy expenditure and on glucose uptake in isolated soleus muscle of obese mice. A seven-day intraperitoneal treatment with the CB_1 receptor antagonist rimonabant caused a 37% increase in basal oxygen consumption compared to that of vehicle-treated animals, and a significant 68% increase in glucose uptake in isolated soleus muscle preparation. To explain this finding, theauthors speculated that rimonabant may act by stimulating efferent sympathetic activity. However, they also hypothesized that the increase in thermogenesis might also be due to an increase in free fatty acid oxidation promoted by the rimonabant-stimulated adiponectin release (Liu et al. 2005). Regardless of the mechanism, the authors concluded that rimonabant has a direct effect on energy expenditure, suggesting that the anti-obesity effect of rimonabant is due to activation of thermogenesis in addition to the initial hypophagia. In addition, the increase in soleus muscle glucose uptake with rimonabant treatment may contribute to the improved glycemia observed in clinical studies (Despres et al. 2005; Van Gaal et al. 2005).

The CB_1 receptor antagonist rimonabant in the treatment of obesity

Obesity is the most common metabolic disease in developed nations. The World Health Organization (WHO) has estimated that, worldwide, more than one billion adults are overweight, with at least 300 million of them being obese (Bays 2004). In addition, the WHO has estimated that, yearly, about a quarter of a million deaths in Europe and more than 2.5 million deaths worldwide are weight-related, with cardiovascular disease as the leading cause. Although the prevalence of the disease is increasing rapidly, the pharmacological treatment options for obesity are currently very limited (York et al. 2004).

Rimonabant is a selective antagonist of CB_1 receptors and is being developed for the potential treatment of obesity and smoking cessation (Carai et al. 2005; Fernandez and Allison 2004). This compound produces weight loss and ameliorates metabolic abnormalities in obese animals (Ravinet-Trillou et al. 2003; Cota et al. 2003b; Black 2004). Results from a large, multicentre, multi-national, randomised, placebo-controlled trial (phase III clinical trial), assessing the efficacy and safety of rimonabant in reducing body weight and improving cardiovascular risk factors in overweight or obese patients, have been recently published (Van Gaal et al. 2005). Overweight patients ($n = 1057$) with treated or untreated dyslipidemia, hypertension or both, were randomised to receive double-blind treatment with placebo, 5 mg rimonabant, or 20 mg rimonabant once daily in addition to a mild hypocaloric diet (600 kcal/day deficit). In this study, treatment with 20 mg rimonabant over one year lead to sustained, clinically meaningful weight loss, reduction in waist circumference, and associated improvements in several cardiovascular and metabolic risk factors, including HDL cholesterol and triglyceride concentrations, and prevalence of the metabolic syndrome. About half of the effect of rimonabant on HDL cholesterol and triglycerides was independent of weight loss. The pattern of weight loss observed in this study with rimonabant appears to be sustained up to 36–40 weeks (Despres et al. 2005). The weight loss observed in 39% of patients treated with 20 mg rimonabant was associated with a concomitant reduction in waist circumference of about 9 cm, a value that could be associated with a 30% decrease in intra-abdominal adiposity (Despres et al. 2001). These results have been recently confirmed by another authoritative trial (Despres et al. 2005).

Rimonabant treatment was well tolerated during this trial, with a similar overall dropout rate in all treatment groups. The most common adverse events experienced with 20 mg rimonabant, such as nausea and diarrhea, were found to be mild and generally occurred in the first few months of the treatment. Gastrointestinal side effects might be explained by the mechanism of action of the drug, since enteric CB_1 receptors are involved in the control of intestinal motility (Coutts and Izzo 2004). Serious adverse events did not seem to occur more frequently in the patients treated with rimonabant than in those on placebo.

In summary, rimonabant has several potential advantages (Fernandez and Allison 2004): 1) it has a good chance to reach the market; 2) it can be administered orally, which gives it a substantial marketing advantage over some other, potential anti-obesity compounds in the pipeline; 3) it is well tolerated; and 4) the rebound of lost weight after discontinuing the drug is less than with other compounds. The strength of its therapeutic effect, however, is not radically different from the modest effects generally observed with other available antiobesity drugs (rimonabant, 20 mg/kg, produces approximately six to seven kilograms decrease in body weight compared to placebo after a one-year treatment). However, rimonabant has a safety profile superior to those of the currently employed antiobesity drugs.

Conclusions

Considerable research has examined endocannabinoid involvement in appetite, eating behavior and body weight regulation. It is now well established that endocannabinoids acting at central (hypothalamus, limbic system) and peripheral (gastrointestinal, liver, skeletal muscles, adipose tissue) CB_1 receptors may positively regulate appetite and energy balance. Conversely, CB_1 cannabinoid receptor antagonists reduce food intake and body weight in rodents, and clinical trials have shown that rimonabant, a selective CB_1 receptor antagonist, reduces body weight and cardiovascular risk factors in overweight patients. Thus, blockade of CB_1 receptors represents a new strategy for treating the obesity that afflicts an increasing number of patients in developed countries.

References

Abel EL (1975) *Cannabis*: effects on hunger and thirst. Behav Biol 15:255–281
Abel EL (1971) Effects of marihuana on the solution of anagrams, memory and appetite. Nature 231:260–261
Arnone M, Maruani J, Chaperon F, Thiebot MH, Poncelet M, Soubrie P, Le Fur G (1997) Selective inhibition of sucrose and ethanol intake by SR 141716, an antagonist of central cannabinoid (CB1) receptors. Psychopharmacology (Berl) 132:104–106
Bays HE (2004) Current and investigational antiobesity agents and obesity therapeutic treatment targets. Obes Res 12:1197–1211
Bensaid M, Gary-Bobo M, Esclangon A, Maffrand JP, Le Fur G, Oury-Donat F, Soubrie P (2003) The cannabinoid CB1 receptor antagonist SR141716 increases Acrp30 mRNA expression in adipose tissue of obese fa/fa rats and in cultured adipocyte cells. Mol Pharmacol 63:908–914
Berry EM, Mechoulam R (2002) Tetrahydrocannabinol and endocannabinoids in feeding and appetite. Pharmacol Ther 95:185–190
Black SC (2004) Cannabinoid receptor antagonists and obesity. Curr Opin Investig Drugs 5:389–394
Boyd ST, Fremming BA (2005) Rimonabant-a selective CB1 antagonist. Ann Pharmacother 39:684–690
Brown JE, Kassouny M, Cross JK (1977) Kinetic studies of food intake and sucrose solution preference by rats treated with low doses of delta9-tetrahydrocannabinol. Behav Biol 20:104–110

Burdyga G, Lal S, Varro A, Dimaline R, Thompson DG, Dockray GJ (2004) Expression of cannabinoid CB1 receptors by vagal afferent neurons is inhibited by cholecystokinin. J Neurosci 24:2708–2715

Carai MA, Colombo G, Gessa GL (2005) Rimonabant: the first therapeutically relevant cannabinoid antagonist. Life Sci 77:2339–2350

Carlini EA, Kramer C (1965) Effects of *Cannabis* sativa (marihuana) on maze performance of the rat. Psychopharmacologia 7:175–181

Chen RZ, Huang RR, Shen CP, MacNeil DJ, Fong TM (2004) Synergistic effects of cannabinoid inverse agonist AM251 and opioid antagonist nalmefene on food intake in mice. Brain Res 999:227–230

Chiesi M, Huppertz C, Hofbauer KG (2001) Pharmacotherapy of obesity: targets and perspectives. Trends Pharmacol Sci 22:247–254

Colombo G, Agabio R, Diaz G, Lobina C, Reali R, Gessa GL (1998) Appetite suppression and weight loss after the cannabinoid antagonist SR 141716. Life Sci 63:PL113–117

Cota D, Marsicano G, Lutz B, Vicennati V, Stalla GK, Pasquali R, Pagotto U (2003a) Endogenous cannabinoid system as a modulator of food intake. Int J Obes Relat Metab Disord 27:289–301

Cota D, Marsicano G, Tschop M, Grubler Y, Flachskamm C, Schubert M, Auer D, Yassouridis A, Thone-Reineke C, Ortmann S, Tomassoni F, Cervino C, Nisoli E, Linthorst AC, Pasquali R, Lutz B, Stalla GK, Pagotto U (2003b) The endogenous cannabinoid system affects energy balance via central orexigenic drive and peripheral lipogenesis. J Clin Invest 112:423–431

Coutts AA, Izzo AA (2004) The gastrointestinal pharmacology of cannabinoids: an update. Curr Opin Pharmacol 4:572–579

Despres JP, Lemieux I, Prud'homme D (2001) Treatment of obesity: need to focus on high risk abdominally obese patients. BMJ 322:716–720

Despres JP, Golay A, Sjostrom L (2005) Rimonabant in Obesity-Lipids Study Group. Effects of rimonabant on metabolic risk factors in overweight patients with dyslipidemia. N Engl J Med 353:2121–2134

Deutsch DG, Ueda N, Yamamoto S (2002) The fatty acid amide hydrolase (FAAH). Prostaglandins Leucotr Essent Fatty Acids 66:201–210

Devane WA, Hanus L, Breuer A, Pertwee RG, Stevenson LA, Griffin G, Gibson D, Mandelbaum A, Etinger A, Mechoulam R (1992) Isolation and structure of a brain constituent that binds to the cannabinoid receptor. Science 258:1946–1949

Devane WA, Dysarz FAI, Johnson MR, Melvin LS, Howlett AC (1988) Determination and characterization of a cannabinoid receptor in rat brain. Mol Pharmacol 34:605–613

Di Girolamo M, Harp J, Stevens J (2000) Obesità: definition and epidemiology. In: Lockwood DH, Heffner TG (eds) Obesità: pathology and therapy. Springer, Berlin, p 4–28

Di Marzo V, Matias I (2005) Endocannabinoid control of food intake and energy balance. NatureNeurosci 8:585–589

Di Marzo V, Goparaju SK, Wang L, Liu J, Batkai S, Jarai Z, Fezza F, Miura GI, Palmiter RD, Sugiura T, Kunos G (2001) Leptin-regulated endocannabinoids are involved in maintaining food intake. Nature 410:822–825

Fernandez JR, Allison DB (2004) Rimonabant Sanofi-Synthelabo. Curr Opin Investig Drugs 5:430–435

Foltin RW, Brady JV, Fischman MW (1986) Behavioral analysis of marijuana effects on food intake in humans. Pharmacol Biochem Behav 25:577–582

Foltin RW, Fischman MW, Byrne MF (1988) Effects of smoked marijuana on food intake and body weight of humans living in a residential laboratory. Appetite 11:1–14

Fowler CH, Jacobsson SOP (2002) Cellular transport of anandamide, 2-arachidonoylglycerol and palmitoylethanolamide – targets for drug development? Prostaglandins Leucotr Essent Fatty Acids 66:193–200

Freedland CS, Poston JS, Porrino LJ (2000) Effects of SR141716A, a central cannabinoid receptor antagonist, on food-maintained responding. Pharmacol Biochem Behav 67:265–270

Fride E (2004) The endocannabinoid-CB receptor system: Importance for development and in pediatric disease. Neuro Endocrinol Lett 25:24–30

Fride E, Bregman T, Kirkham TC (2005) Endocannabinoids and food intake: newborn suckling and appetite regulation in adulthood. Exp Biol Med (Maywood). 230:225–234

Giuffrida A, Beltramo B, Piomelli D (2001) Mechanisms of endocannabinoid inactivation: biochemistry and pharmacology. J Pharmacol Exp Ther 298:7–14

Gomez R, Navarro M, Ferrer B, Trigo JM, Bilbao A, Del Arco I, Cippitelli A, Nava F, Piomelli D, Rodriguez de Fonseca F (2002) A peripheral mechanism for CB1 cannabinoid receptor-dependent modulation of feeding. J Neurosci. 22:9612–9617

Greenberg I, Kuehnle J, Mendelson JH, Bernstein JG (1976) Effects of marihuana use on body weight and caloric intake in humans. Psychopharmacology 49:79–84

Hanus L, Abu-Lafi S, Fride E, Breuer A, Vogel Z, Shalev DE, Kustanovich I, Mechoulam R (2001) 2-arachidonyl glyceryl ether, an endogenous agonist of the cannabinoid CB_1 receptor. Proc Natl Acad Sci USA 98:3662–3665

Hanus L, Avraham Y, Ben-Shushan D, Zolotarev O, Berry EM, Mechoulam R (2003) Short-term fasting and prolonged semistarvation have opposite effects on 2-AG levels in mouse brain. Brain Res 983:144–151

Hilairet S, Bouaboula M, Carriere D, Le Fur G, Casellas P (2003) Hypersensitization of the Orexin 1 receptor by the CB1 receptor: evidence for cross-talk blocked by the specific CB1 antagonist, SR141716. J Biol Chem 278:23731–23737

Hildebrandt AL, Kelly-Sullivan DM, Black SC (2003) Antiobesity effects of chronic cannabinoid CB_1 receptor antagonist treatment in diet-induced obese mice. Eur J Pharmacol 462:125–132

Hollister LE (1971) Hunger and appetite after single doses of marihuana, alcohol, and dextroamphetamine. Clin Pharmacol Ther 12:44–49

Howlett AC, Barth F, Bonner TI, Cabral G, Casellas P, Devane WA, Felder CC, Herkenham M, Mackie K, Martin BR, Mechoulam R, Pertwee RG (2002) International Union of Pharmacology. XXVII. Classification of cannabinoid receptors. Pharmacol Rev 54:161–202

Jamshidi N, Taylor DA (2001) Anandamide administration into the ventromedial hypothalamus stimulates appetite in rats. Br J Pharmacol 134:1151–1154

Kirkham TC (1990) Enhanced anorectic potency of naloxone in rats sham feeding 30% sucrose: reversal by repeated naloxone administration. Physiol Behav 47:419–426

Kirkham TC, Williams CM (2001) Synergistic efects of opioid and cannabinoid antagonists on food intake. Psychopharmacology (Berl) 153:267–270

Kirkham TC, Williams CM (2004) Endocannabinoid receptor antagonists: potential for obesity treatment. Treat Endocrinol 3:345–360

Kirkham TC, Williams CM, Fezza F, Di Marzo V (2002) Endocannabinoid levels in rat limbic forebrain and hypothalamus in relation to fasting, feeding and satiation: stimulation of eating by 2-arachidonoyl glycerol. Br J Pharmacol 136:550–557

Lichtman AH, Cravatt BF (2005) Food for thought: endocannabinoid modulation of lipogenesis. J Clin Invest 115:1130–1133

Liu YL, Connoley IP, Wilson CA, Stock MJ (2005) Effects of the cannabinoid CB1 receptor antagonist SR141716 on oxygen consumption and soleus muscle glucose uptake in Lep(ob)/Lep(ob) mice. Int J Obes (Lond) 29:183–187

Matsuda LA, Lolait SJ, Brownstein BJ, Youg AC, Bonner TL (1990) Structure of a cannabinoid receptor and functional expression of the cloned cDNA. Nature 346:561–564

Mechoulam R, Ben-Shabat S, Hanus L, Ligumsky M, Kaminski NE, Schatz AR, Gopher A, Almog S, Martin BR, Compton DR, Pertwee R, Griffin G, Bayewitch M, Barg J, Vogel Z (1995) Identification of an endogenous 2-monoglyceride, present in canine gut, that binds to a cannabinoid receptor. Biochem Pharmacol 50:83–90

Munro S, Thomas KL, Abu-Shaar M (1993) Molecular characterization of a peripheral receptor for cannabinoids. Nature 365:61–65

Nocerino E, Amato M, Izzo AA (2000) *Cannabis* and cannabinoid receptors. Fitoterapia 71:S6–S12

Osei-Hyiaman D, DePetrillo M, Pacher P, Liu J, Radeva S, Batkai S, Harvey White J, Mackie K, Offertaler L, Wang L Kunos G (2005) Endocannabinoid activation at hepatic CB1 receptors stimulates fatty acid synthesis and contributes to diet-induced obesity. J Clin Invest 115:1298–1305

Pagotto U, Vicennati V, Pasquali R (2005) The endocannabinoid system and the treatment of obesity. Ann Med 37:270–275

Poncelet M, Maruani J, Calassi R, Soubrie P (2003) Overeating, alcohol and sucrose consumption decrease in CB1 receptor deleted mice. Neurosci Lett 343:216–218

Ravinet Trillou C, Arnone M, Delgorge C, Gonalons N, Keane P, Maffrand JP, Soubrie P (2003) Anti-obesity effect of SR141716, a CB1 receptor antagonist, in diet-induced obese mice. Am J Physiol Regul Integr Comp Physiol 284:R345–353

Regelson WDR, Boyle LE, Smith TL, Gehan EA, Samuels ML (1976) Delta-9-tetrahydrocannabinol as an effective antidepressant and appetite-stimulant agent in advanced cancer patients. In: Brande MC, Szara S (eds) The pharmacology of marijuana. Raven Press: New York pp 763–766

Rowland NE, Mukherjee M, Robertson K (2001) Effects of the cannabinoid receptor antagonist SR 141716, alone and in combination with dexfenfluramine or naloxone, on food intake in rats. Psychopharmacology (Berl) 159:111–116

Siler JF, Sheep WL, Bates LB, Clark GF Cook GW, Smith WH (1933) Marihuana smoking in Panama. Mil Surg 73:269–280

Simiand J, Keane M, Keane PE, Soubrie P (1998) SR 141716, a CB1 cannabinoid receptor antagonist, selectively reduces sweet food intake in marmoset. Behav Pharmacol 9:179–181

Sugiura T, Kondo S, Sukagawa A, Nakane S, Shinoda A, Itoh K, Yamashita A, Waku K (1995) 2-arachidonoylglycerol: a possible endogenous cannabinoid receptor ligand in brain. Biochem Biophys Res Commun 215:89–97

Sugiura T, Kobayashi Y, Oka S, Waku K (2002) Biosynthesis and degradation of anandamide and 2-arachidonylglycerol and their possible physiological significance. Prostaglandins Leucotr Essent Fatty Acids 66:173–192

Touw M (1981) The religious and medicinal uses of *Cannabis* in China, India and Tibet. J Psychoactive Drugs 13:23–34

Tucci SA, Rogers EK, Korbonits M, Kirkham TC (2004) The cannabinoid CB1 receptor antagonist SR141716 blocks the orexigenic effects of intrahypothalamic ghrelin. Br J Pharmacol 143:520–523

Van Gaal LF, Rissanen AM, Scheen AJ, Ziegler O, Rossner S (2005) RIO-Europe Study Group. Effects of the cannabinoid-1 receptor blocker rimonabant on weight reduction and cardiovascular risk factors in overweight patients: 1-year experience from the RIO-Europe study. Lancet 365:1389–1397

Vickers SP, Kennett GA (2005) Cannabinoids and the regulation of ingestive behaviour. Curr Drug Targets 6:215–223

Williams CM, Kirkham TC (1999) Anandamide induces overeating: mediation by central cannabinoid (CB1) receptors. Psychopharmacology (Berl) 143:315–317

Williams CM, Rogers PJ, Kirkham TC (1998) Hyperphagia in pre-fed rats following oral delta9-THC. Physiol Behav 65:343–346

Williamson EM, Evans FJ (2000) Cannabinoids in clinical practice. Drugs 60:1303–1314

Wilson RI, Nicoll RA (2002) Endocannainoid signaling in the brain. Science 296:678–682

York DA, Rossner S, Caterson I, Chen CM, James WP, Kumanyika S, Martorell R, Vorster HH (2004) American Heart Association. Prevention Conference VII: Obesity, a worldwide epidemic related to heart disease and stroke: Group I: worldwide demographics of obesity. Circulation 110:e463–470

Zygmunt PM, Petersson J, Andersson DA, Chuang H, Sorgard M, Di Marzo V, Julius D, Hogestatt ED (1999) Vanilloid receptors on sensory nerves mediate the vasodilatator action of anandamide. Nature 400:452–457

Subject Index

Printing: Krips bv, Meppel
Binding: Stürtz, Würzburg